Digital Media Ethics

Digital Media and Society Series

New technologies are fundamentally altering the ways in which we communicate. This series from Polity aims to provide a set of books that make available to a broad readership cutting-edge research and thinking on digital media and their social contexts. Taken as a whole, the series will examine questions about the impact of network technology and digital media on society in all its facets, including economics, culture and politics.

Jean Burgess and Joshua Green: *YouTube*
Mark Deuze: *Media Work*
Charles Ess: *Digital Media Ethics*
Alexander Halavais: *Search Engine Society*
Robert Hassan: *The Information Society*
Tim Jordan: *Hacking*
Rich Ling and Jonathan Donner: *Mobile Communication*
Donald Matheson and Stuart Allan: *Digital War Reporting*
Jill Walker Rettberg: *Blogging*
Patrik Wikström: *The Music Industry*

Digital Media Ethics

CHARLES ESS

polity

First published in 2009 by Polity Press

Reprinted in 2010

Polity Press
65 Bridge Street
Cambridge CB2 1UR, UK

Polity Press
350 Main Street
Malden, MA 02148, USA

ISBN-13: 978-0-7456-4163-8
ISBN-13: 978-0-7456-4164-5 (pb)

A catalogue record for this book is available from the British Library.

Typeset in 10.25 on 12.5 pt FF Scala
by Servis Filmsetting Ltd, Stockport, Cheshire
Printed and bound in the United States by Odyssey Press Inc.,
Gonic, New Hampshire

Text design by Peter Ducker MISTD

For further information on Polity, visit our website: www.politybooks.com

For Mom (1927–2008) and Dad, my first and most important ethics teachers.

For Conni, best beloved, wonderful mother, grace-ful minister.

For Joshua and Kathleen, our children, now young adults, who bring overwhelming joy and hope.

Contents

Foreword

LUCIANO FLORIDI

A common risk, run by many Forewords, is to bother the reader by repeating, sometimes less accurately, what the table of contents of the book already specifies, or (and unfortunately this is often an inclusive 'or') by eulogizing the text and the author, plastering comments that look like semantic clones lifted from a myriad of other texts. It is in order to try to avoid both pitfalls that I shall skip here the usual hypes – which the book and its author do deserve, make no mistake – in order to speak to the reader a bit more frankly and hence, I hope, less uninformatively.

The book has all the usual virtues of a good textbook: it is carefully researched, clearly written, and argued intelligently. Yet these are basic features that we have come to expect from high-standard scholarship and do not make it special. That Charles Ess has written a good textbook is uninteresting. That he might have written an excellent one is what I would like to argue.

What the book offers, over and above its competitors, are some remarkable and, to my knowledge, unique features. Let me be schematic. The list is not exhaustive, nor do the listed features appear in order of importance, but there is a good narrative that keeps them together.

First, the topic. The book addresses the gray but crucial area of ethical concerns raised by digital media. Of course, it is flanked on the shelf by many other textbooks in Information and Computer Ethics but, as Charles Ess well explains, this is not one of them, and sticks out for its originality. For the book tackles that messy area of our ordinary lives where ethical issues are entangled with digital mass media, communication artifacts, information technologies of all sorts, computational processes, computer-mediated social interactions and so forth. Indeed, it is one of its virtues that it tries to clarify that "so forth" which I have just somewhat surreptitiously added in order to spare myself the embarrassment of a lack of a clear definition. As Schrödinger once said in a different context, this is a very sharp picture of a rather fuzzy subject.

Second, the approach. The book has all the required philosophical rigor but, once again, this is not its most impressive feature. It is also graced by a light touch, which means that Ess has avoided being either *prescriptive* or *proscriptive* (you will not be told what to do and what not to do), opting in favor of an enlightened (liberal, in his own words), *critical description* of the problems discussed. This is a noteworthy advantage, since the author empowers the reader, as should be (but often is not) the case with similar texts. Having said all this, the feature that I find absolutely unique and outstanding (in the sense that it makes this book stand out on the ideal shelf of other comparable books) is its capacity to combine a pluralistic approach – without the bitter aftertaste of some crypto-relativism – with a well-informed and timely look into non-Western views on the ethical issues it tackles. This is crucial. Following a remarkable tradition of German philosophers (Nietzsche, Schopenhauer, Hegel), Ess makes a sustained and successful effort to bring together Eastern and Western ethical traditions in an enriching and fascinating synthesis. And he achieves all this thanks to his extended, international experiences with a variety of cultures. If you wish to see how masterfully Ess avoids syncretism, relativism, and dogmatism and succeeds in shaping an overview of the field which is both captivating and ethically robust, you need to read the book.

Third, the style. This is a reader-friendly book that teaches without patronizing, with a didactic style that can only be the result of decades of care and experience in guiding students and readers through difficult topics. Its degree of accessibility is as misleading as the ability of an acrobat to make his or her performance look effortless.

Someone once told me that digital media are like pornography: it is very difficult to define them, but you recognize them immediately when you see them. Because we all know what digital media are, even if it is hard to determine the boundaries of their nature, applications, evolutions, and effects on our lives, I am confident that the reader will understand why I would recommend this book not only inside but also outside the classroom. Given its topic, its approach and its style, this is a book for the educated public as well. It should be read by anyone interested in the development and future of the information society and our moral lives within it.

Preface
Why This Book?

There are many excellent textbooks in applied ethics as well as in Information and Computing Ethics (ICE). Indeed, I draw upon and refer to these throughout this text. But here I try to do three things a bit differently than in other textbooks. First, I try to address here a wider audience, and, thereby, a correlatively wider range of issues and problems, than those characteristically associated with ICE. Because I think that digital media ethics are important to anyone who uses a computer, a cellphone, videogames, and so forth, this text is designed for use in a wide range of classes, including media, communication, and information classes (e.g., related to computer science, librarianship, etc.), not solely classes in philosophy and applied ethics. To be sure, we explore here some of the most important problems and issues in ICE – privacy, copyright, pornography, as well as freedom of expression and cross-cultural communication online. But I leave out a number of additional and important topics (e.g., security, hacking, and the specific obligations of professional ethics) in favor of further including digital media such as cellphones and videogames and some of the ethical issues they likewise entail.

Second, I don't try to be comprehensive. Rather, my simple thought is that less is sometimes more. Experience has taught me that this is often true in effective teaching, and this volume is designed primarily as a teaching text; this will be apparent from the outset in the manifold case-studies used in the beginning and then throughout each chapter, each accompanied by questions designed to foster individual and collaborative reflection, discussion, writing, and research.

In this way, the book is intended to serve as primer. The goal is to provide students and instructors in a wide range of disciplines with a first introduction to important ethical issues associated with digital media (chapters 1–5). This is conjoined with an initial exposure to the ethical and meta-ethical frameworks that are commonly used in applied ethics (chapter 6). I'm acutely aware that my introductions and discussions do not (usually) go into the level of detail, including important historical and conceptual backgrounds, that is necessary for more robust analyses and philosophical reflection. To be sure, each chapter includes more fine-grained reflection and research exercises and relevant resources intended to help students and instructors pursue these matters more fully. I hope that these – along with the further materials and research that instructors and their classes will want to explore – will thereby help readers develop more nuanced understandings and insights that build on necessarily basic introductions.

The book is driven by a basic recognition: given the rapid development and expansion of these media, within both developed and developing countries, it seems certain that more or less *all of us* – not solely those of us privileged to pursue philosophy – will only continue to encounter more and more new sorts of ethical issues and conundrums. By starting, in effect, with the basics, I hope that those who use this book will learn how to grapple effectively with contemporary and prevailing ethical issues, and thereby be better prepared to take on more effectively those new issues that will inevitably arise in the future.

The third difference between this and other textbooks is that this approach drives its primary pedagogical orientation and principles. Reflecting especially my commitments to liberal arts education (education for *free* – *liber* – people), I'm not here in the first instance to tell anyone what to think. Rather, I have tried to focus on the ethical issues as genuine questions – i.e., questions that admit of *multiple*, ethically justifiable responses, not (usually) a single right one. (As we will see, this approach is affiliated with the meta-ethical position of *pluralism*; see chapter 6 for further discussion.) To be sure, I certainly have some opinions about these issues; I will try to make these clear, so that readers are aware of my opinions and biases. But I hope what will be immediately striking is the effort to provide conceptual frameworks, relevant information, and sufficient examples to help you practice *applying* those frameworks to specific ethical issues. As I often repeat to my own students: I don't care *what* you think (short of, say, affirming Neo-nazi racism, at least), but I very much care *how* you think, i.e., whether or not you can articulate and defend with sound logic and shared empathies your own views, while simultaneously understanding the views and arguments of those who disagree with you. I hope that, in exploring the diverse ethical frameworks that both we *and* Others use, we will become more informed and articulate regarding the ethical frameworks and larger cultural norms, traditions, and practices that shape both how we and Others draw our ethical conclusions.

Learning how to do so is critical, for a number of reasons. Some of these are prudential or self-interested. Again, it seems a given that more and more of us will encounter more and more ethical issues evoked by digital media; learning how to deal with these sorts of issues will hence better prepare us to respond in ethically informed and responsible ways to what is to come. Moreover, there is good evidence to suggest that as we learn to become more reflective and aware of our primary ethical commitments, we are thereby better able to "do the right thing" when the time comes. For example, as the famous Milgram experiments demonstrated, those best able to resist or refuse to follow orders by authority figures in lab coats are those who are most reflective and aware of why they hold to certain basic ethical principles or values.

Finally, in keeping with the liberal arts commitments and virtue ethics traditions (both Western and non-Western) represented here, developing our capacities for ethical reflection is central in two deeply *human* ways. First, we come to better understand our own ethical frameworks, alongside our own cultural norms and practices; but this also helps us better understand the frameworks, norms, and practices of Others – especially as these often lead to different, sometimes conflicting conclusions as to what the right thing to do may be in a particular instance. In other words, we better learn to understand the perspective of "the Other," even (especially) when we disagree. As we will see, these disagreements are often rooted in the different ethical norms, traditions, and practices associated with specific cultures. And so I argue in chapter 4 that our learning to better understand the perspective of "the Other," especially across different cultures, is increasingly essential for us as citizens of a *global* society – one ever more interwoven by the digital media we examine in this text.

At the same time, this deeper understanding of "the Other" does *not* mean we necessarily must change our own minds (or cultures). But, ideally, I hope this effort towards perspective-taking will lead to the experience that often we learn the most from those with whom we *dis*agree. By opening ourselves up to the perspectives, experience, and cultures of the many Others around us, we often gain insights and understanding we would otherwise not acquire. Such understanding, moreover, is not only self-enriching; it is further crucial to our learning how to get along with one another in an increasingly interconnected world. Learning more about the perspectives and cultures of Others will not only help us move towards a more genuinely welcoming posture with regard to those who are (sometimes profoundly) different from us; our resulting understandings may further suggest how our differences can be resolved in more harmonious rather than conflict-driven ways.

A second central benefit of enhancing our capacities for ethical reflection is the development of what Socrates and Aristotle highlighted as *phronesis*: our ability to make well-informed *judgments* in specific and often novel contexts. The ability to judge in these ways is not only necessary if we are to make sound and responsible decisions; as philosophers and religious teachers in many traditions, East and West, have argued, such judgment is further central to our becoming more excellent human beings – i.e., human beings who practice and learn how to be *good* human beings, human beings good at practicing such human excellences or virtues as care and compassion, understanding, charity, forgiveness, and so forth. As we will explore more fully, the pursuit of such excellences is its own reward: philosophers and religious teachers, both West and East, have discerned that the disciplined pursuit and practice of ethical judgment and its affiliated virtues lead to a deep harmony – both a self-harmony and harmony with the larger community. Such harmony is felt in terms of a deep contentment and satisfaction with our lives, both as individuals and as members of larger communities.

Just to be clear: this is one of the viewpoints I'm personally most attracted to. Hence, it is at work throughout this volume. Indeed, as I occasionally suggest, pursuing these practices and excellences is no longer a luxury for a privileged few. Socrates, Aristotle, and Confucius presumed that these sorts of reflections and efforts to become an *excellent* human being were possible only for an intellectual and social elite – the "few," not "the many." But modern liberal democratic states are built on the assumption that "the many" are also capable of such understanding and ethical judgment. These are necessary, that is, if we are to regulate our own affairs and to engage with others in ways not defined solely by force and power – rather than being regulated instead by some sort of authoritarian regime. Such understanding and ethical judgment, in short, is necessary if we are to become and pursue our lives as free (in both ethical and political senses) human beings. Moreover, whatever sort of nation-state we may inhabit, in a world of networked digital media that increasingly interconnects our lives in ever-expanding webs of relationships with others throughout the diverse cultures of the globe, like it or not, we are all increasingly *cosmopolitans*, citizens of the world, not simply citizens of a given nation. For us as *cosmopolitans*, building and refining our capacity for responsible ethical judgments, as informed by a deep understanding of the ethical frameworks and cultural norms that shape those judgments in both ourselves and others, is a necessity.

I very much hope the material and exercises collected here in fact help in that endeavor.

Acknowledgements

As is mentioned frequently in this book, the rise of computer networks is leading many of us in the West to recognize something called "distributed responsibility" – the understanding that as these networks make us more and more interwoven with and interdependent upon one another, the ethical responsibility for a given act is distributed across a network of actors, not simply attached to a single individual. At the same time, this awareness of our absolute interdependence on others is ancient, as premodern ethics, both East and West, emphasize the importance of community. For me, this awareness is rarely more acute than when trying to properly recognize the numerous persons who have contributed to this volume. The web of relationships brought to a focus in this book begins with family, friends, my own teachers and colleagues – including John Lawrence, my first guru in the then (1980s) nascent field of humanities computing; Preston Covey (1942–2006), whose pioneering work and generous friendship in computing and applied ethics inspired and encouraged so many of us in these fields; and Henry Rosemont, Jr., whose patience in conveying the elements of Confucian thought has been exemplary indeed.

I am further very grateful to Andrea Drugan at Polity Press for initially suggesting the idea of putting together a book like this. She and her colleague Jonathan Skerrett have been consistently supportive and patient throughout the process. In particular, Andrea is an ideal editor: her sharp eye and excellent suggestions have improved this text throughout. I am equally grateful to two initially anonymous reviewers whose critically constructive comments on an early version of the text were extremely useful. I am delighted that I can now thank them by name for their help. Richard Volkman (University of Southern Connecticut) and Herman Tavani (Rivier College) pointed out embarrassing errors and shortcomings, made many excellent suggestions for improvement, and offered encouraging support for the bits they thought I got right. So far as I've managed to respond effectively to their critiques, this is a much better text.

Numerous friends and colleagues have offered helpful advice and suggestions along the way. It is a particular pleasure to thank my colleagues at the Institut for Informations- og Medievidenskab (Institute of Information and Media Studies), University of Aarhus, Denmark, beginning with Jakob Linaa Jensen, Niels Ole Finnemann, Niels Brügger, Finn Olesen, Randi Markussen, Poul Erik Nielsen, and Institute Head Steffen Ejnar Brandorff. The Danish flavor of this text – most apparent in some of the examples and case-studies – reflects the very productive and enjoyable fall semester, 2007, I spent at the Institute as a Visiting Professor. The research and writing that this stay made possible, along with my Danish colleagues' generous support and suggestions, directly contributed to the first stages of this text. My Drury philosophy colleagues – Dr. Lisa Esposito, Dr. Chris Panza, and Dr. Ted Vaggalis – likewise provided invaluable insight and encouragement. Finally, Ms. Erin Marie Tracy, our fabulous student assistant in the Interdisciplinary Studies Center, diligently worked through the entire manuscript, pointing out especially where student readers, however intelligent and well motivated, would likely be derailed by my overuse of dashes and German-length sentences.

Several others have been particularly helpful with specific issues and chapters. Dr. Heather Johnson (Stevens Technical Institute) provided especially detailed criticisms and suggestions for chapters 2 and 3 on privacy; I am also grateful to her students for helping to "road test" portions of the text, including some of the pedagogical materials intended precisely for classroom and student use. Dr. Richard Schur (Drury University) was particularly helpful with chapter 3, bringing to bear his considerable expertise in matters of copyright law. Dr. Susanna Paasonen (University of Helsinki) provided invaluable guidance through the complicated and often murky waters of pornography (chapter 5). Dr. Mia Consalvo (Ohio

University) and Dr. Miguel Sicart (I.T.-University, Copenhagen) were equally generous and helpful with the section on games (chapter 5).

Finally, my family played the most important roles – first of all, as their unending love and support make all of my work and play possible, meaningful, and occasionally even fun. My brother Robert (Director, I.T. Operations, Fujitsu Network Communications) provided most helpful technical insight as well as a professional's perspective. My wife, the Reverend Conni Ess, put up with the brunt of the workaholic practices and schedules needed to bring a book like this to fruition. Our son Joshua, a computer geek now turned professional technical support person, helped especially with the more arcane technical details that often prove essential to digital media ethics. Our daughter Kathleen, pursuing her own academic career as a classics and religion scholar, provided very helpful assistance with both Greek philosophy and English style. My parents, Bob and Betty Ess, are to be thanked for the innumerable ways they fostered and encouraged this work.

My mom was especially pleased to see me working on this project. While her death on May 9, 2008 was in almost every conceivable way a good death, it's a regret that she could not stay with us long enough to see this book in its final form. Like any mother, she was always pleased and proud of her children's accomplishments – especially those that sought to help others. At the same time, in many ways she was the single person most responsible for my pursuing philosophy. I hope this book lives up to her spirit of enacting deep care for others. Insofar as it does, Mom, this one's for you.

Chapter Synopses

Chapter 1: Central Issues in the Ethics of Digital Media

This chapter introduces us to some of the characteristics of digital media that give rise to both familiar and novel ethical challenges. We further examine in an initial way how popular media report in the form of "moral panics" – i.e., stories that highlight apparent risks and dangers of digital media; such reporting may entail ways of thinking that can obstruct or short-circuit the sorts of ethical reflection needed to resolve ethical problems more effectively. The chapter concludes with an overview of the organization of the book and its chief pedagogical approaches and apparatus.

Chapter 2: Privacy in the Electronic Global Metropolis?

This chapter addresses the challenges to modern Western notions of individual privacy presented by contemporary digital media, most especially networked communications media such as the Internet and the Web. It then explores diverse *cultural* attitudes towards and understandings of individual and collective privacy in non-Western cultures, in order to raise a central problem for contemporary digital media ethics – namely, is it possible to develop ethical frameworks for the use of digital media that conjoin norms shared globally along with the irreducible differences that define distinctive individual and cultural identities? The meta-theoretical approach of *ethical pluralism* is highlighted here, as contemporary intercultural dialogues show how such ethical pluralism resolves this central problem.

Chapter 3: Copying and Distributing via Digital Media: Copyright, Copyleft, Global Perspectives

Digital media make copying and distributing information – in the form of music, videos, or texts – far easier than with previous media. We explore some of the common ethical aspects of copying and distributing music and other forms through three main frameworks for interpreting and regulating these as intellectual property: current copyright law (especially in the U.S. and E.U.), various "copyleft" schemes, and Confucian tradition as shaping many non-Western approaches and attitudes.

Chapter 4: Citizenship in the Global Metropolis

The example of the "Muhammad cartoons" illustrates that contemporary digital media dramatically extend the scope and thus the ethically relevant consequences of our actions. We examine what ethical obligations we may have to avoid ethnocentrism (the assumption that our specific cultural norms, beliefs, and practices are somehow universal) and thereby cultural imperialism; how such cultural imperialism may be embedded in the technologies of digital communication media themselves; and what ethical guidelines there may be for cross-cultural communication online that seeks to respect and foster "the Otherness of the Other" – i.e., that seeks to recognize and respect the irreducible differences that define distinctive individual and cultural identities. The ethical frameworks of *virtue ethics* are emphasized here.

Chapter 5: Still More Ethical Issues: Digital Sex and Games

This chapter takes up two of the most likely topics of digital media ethics: pornography online and violence in videogames. It first expands on the discussion in chapter 1 on popular media and moral panics, in order to more extensively clarify how effective ethical reflection often requires us to move beyond the polarization and dualistic thinking characteristic of popular media accounts. We then examine the phenomenon of pornography online – in part, as a reflection, perhaps, of a larger "pornification" of society. *Feminist ethics* and *post-feminist frameworks* are emphasized in the analysis of online pornography. *Virtue ethics* are emphasized in the analysis of violence and videogames.

Chapter 6: Digital Media Ethics: Overview, Frameworks, Resources

This chapter introduces and illustrates by way of several examples the frameworks applied in the topics chapters (2–5). These include the widely used *ethical frameworks* of *utilitarianism, deontology, feminist ethics/ethics of care, virtue ethics,* and *Confucian* and *African* frameworks as prescribing distinctive approaches to analyzing and resolving ethical difficulties. Moreover, three major *meta-theoretical* frameworks – those of *ethical relativism, ethical monism/absolutism,* and *ethical pluralism* – are likewise introduced and examined with some care.

CHAPTER I

Central Issues in the Ethics of Digital Media

Chapter overview

We begin with a case-study intended to introduce us to *privacy* as one of the most significant ethical issues brought about by digital media. This case-study is accompanied by one of the primary *ped-agogical*/teaching elements of the book – questions designed to foster initial reflection and discussion (either for individuals, small groups, or a class at large), followed by additional questions that can be used for further reflection and writing.

Following an introduction to the main body of the chapter, the section "(Ethical) Life in the digital age?" provides a first overview of digital media and their ethical dimensions. I also high-light how more popular treatments of these, however, can become counterproductive to clear and careful ethical reflection. We then turn to some of the specific characteristics of digital media – convergence, digital information as "greased," and digital media as communication technologies – that highlight spe-cific ethical issues treated in this volume. We then take up initial considerations on how to "do" ethics in the age of digital media. Finally, I describe the pedagogical features of the book and provide some suggestions for how it is designed to be used – including specific suggestions for the *order* in which the chapters may be taken up.

Case-study: Facebook and Beacon – an introduction to issues in digital media ethics

In October 2007, the increasingly popular (for many, essential) social net-working site Facebook introduced a new service, called Beacon. Following earlier software enhancements that automatically provided updates on various elements of the Facebook profiles of one's Facebook friends,

Beacon streamed into one's homepage the recent purchases such friends had made – thus serving as a form of advertising for the products involved.

Facebook sold potential advertisers on Beacon in part by assuring them that Facebook users would be able to "opt-in" to the service – i.e., that they would have to explicitly agree to join the service. (This is in contrast with "opt-out" approaches that presume participation by default, and require the individual to explicitly initiate his or her removal from a given service, etc.)

In fact, Facebook only provided its users with notifications that Facebook would broadcast their purchases to their friends (a) when they made such a purchase on an external website and (b) when they returned from the external site to their Facebook account. When users generally ignored these, Facebook presumed that users had thereby consented to Beacon distributing information about their purchases to their friends.

After massive protest, Facebook changed its policy: if users ignored the warnings, Facebook will assume they are now saying "no" to Beacon. At the same time, however, Facebook will not offer the possibility of an opt-out from Beacon. (See Story 2007.)

Questions for discussion

1. Why might users object to Facebook initiating the Beacon program? On what grounds – i.e., what values, principles, and/or other reasons might support these objections?

2. Is Facebook's new policy – i.e., interpreting users' ignoring of their notifications (that their purchase information will be broadcast to their friends' homepages) as a "no" to this use of Beacon – a sufficient response to the objections you developed in (1)? Why and/or why not – i.e., what are your *reasons* and/or other grounds (including feelings, intuition, etc.) for your position here?

Additional reflection, informal writing exercises

1. Review the "Terms of Use" agreement that every Facebook user must agree to before they are allowed to set up a homepage on Facebook. (You can get to this by clicking on the small link "terms.")

   Do you spot anything here that, upon reflection, you might *not* agree with? If so, identify this point, and explain as best you can your *reasons* (and/or other grounds, including feelings, intuition, etc.) for disagreeing with Facebook on this point.

2. Review Facebook's "Content Code of Conduct" (linked to from the "Terms of Use" document).

   Are there elements of this code that you might *not* agree with? Why not?

   Identify this element(s) carefully, and then:

   A. Articulate as clearly as you can (i) what Facebook requires that you disagree with, and (ii) what your own ethical claim/position/view might be as an alternative to Facebook's.
   B. Explain as fully as you can *why* you hold the ethical claim/position/view that you articulate in (A) (ii), above – i.e., what grounds, reasons, values, norms, etc. would you appeal to in order to support your ethical position/claim?
   C. Articulate as clearly as you can the grounds, reasons, values, norms, etc., that might lie behind Facebook's claim/position/view (*perspective-taking*).
   D. Given what are now two different *arguments* – i.e., your argument for your particular ethical view, and the argument that you have reconstructed in (C) – respond to the conflict or disagreement between these two arguments and views.

3. If you already use Facebook or a similar social networking site, did you read the "Terms of Use" and related policies prior to clicking on the "I agree" box?

   If so, why? If not, why not?

   Do you think most people *don't* read such "Terms of Use" – along with, say, the End User License Agreement (EULA) that users must agree to when installing new software? If not, why not?

4. Do you have any additional comments or observations on what you see as some of the ethical issues that emerge here?

Reflection/writing suggestions: media log

Develop a log – a description of what digital device you use, for how long, and for what purpose – for a given period (e.g., 24 hours/3 days/1 week).

Writing – class-discussion questions:

A. As you review this log, any surprises – e.g., are you using a device(s) significantly more or less than you might have originally thought?

B. Can you determine *how* you're using these devices – specifically, what kinds of communication are you using them for?

C. Are there any obvious *ethical* aspects and/or problems that you notice and/or encounter in your use of these technologies? Identify these as carefully as you can.

Class discussion: Collect the ethical aspects/problems of your uses of digital media onto a commonly shared list. These will be useful to refer to as examples as you work through this text – both as some of these are likely to be discussed in the following chapters, and as some may *not* be explicitly covered here. The latter will be especially interesting for further individual and class analysis, discussion, and writing – particularly as you and your colleagues develop greater skill and ability in grappling with such ethical issues and challenges.

D. How does your use of digital media compare with:

(i)   your older and/or younger siblings (if you have any)?
(ii) your parents and/or parents' friends?
(iii) your grandparents and/or grandparents' friends?

This question asks you to consider the *demographics* of digital media use – on the presumption that there will be striking *differences*, e.g., between the usage rates of adolescents and 20-somethings vis-à-vis[1] people in their 40s and 50s, and then people in their 60s and 70s (and older).

---

1 Vis-à-vis: "face-to-face" (French). I use this phrase a lot, e.g. ". . . very different understandings of the role of the state vis-à-vis the life of the individual." It may be tempting to say something like "versus" the life of the individual – but I intend to avoid precisely the presumption of *opposition* or conflict that "versus" implies. Rather, I intend "vis-à-vis" as referring to the whole range of possible relationships (e.g., as equals, as superior/inferior, as complimentary, as oppositional, etc.) between any two things that we may want to think about together. I thereby want to

E. If there are such differences, what are your thoughts about the possible *ethical* implications of these differences?

(For example, we will explore in chapter 3 a suggestion that – at least some – younger people have ethical sensibilities very different from – at least some of – their elders regarding matters of copying digital media, including [illegal] downloading.)

Does it seem that our *ethics* indeed change from generation to generation? And if so, what does that imply regarding claims that ethics should be universal, i.e., applicable in all times and places to all people?

[*Hint*: we will take this up later on in terms of the *meta-ethical* positions of ethical *absolutism*, *relativism*, and *pluralism* – see chapter 6 for more discussion.]

Additional reflection suggestions: digital media and face-to-face communication

1. It is often claimed and observed that online communication, because it offers *anonymity*, encourages greater openness and honesty than most face-to-face (F2F) communication. On the one hand, it is hoped that such freedom of expression will encourage, for example, subordinates in an organization to more openly express their views (including criticisms): presuming that these views and criticisms are then heard by their superiors, the thought is that this leads to a "flattening" of the traditional hierarchies.

On the other hand, anonymity also encourages less attractive forms of communication, including flaming, trolling, cyberstalking, and cyber-bullying.

In your view, do you think/feel that these new possibilities of online communication have led to greater good than harm – or vice versa? Please be ready to justify your response with one or more examples, arguments, etc.

---

insist that we cannot assume we already know what these relationships may be. Rather, part of our work includes attempting to discern just what these relationships may be in fact.

2. It is commonly observed that the recent popularity of text-messaging (SMS) among young people in the United States is in part because we can use SMS to *avoid* face-to-face communication – e.g., some people have been known to break up with a girlfriend or boyfriend using a text-message.

Do you notice ways in which new digital communication technologies, while *increasing* our capabilities to communicate with one another in various ways, somewhat paradoxically thereby also make it possible to *reduce* our communication with one another, e.g., by not answering a phone call from a caller identified by name and number on our phone, by text-messaging instead of discussing important matters face to face, etc.?

Introduction

It is now commonplace to observe that in the industrialized world, we – especially young people – are thoroughly interconnected through, and thereby saturated with, what are sometimes called "New Media" or digital media. (We will explore the meaning of these terms more carefully in a bit.) Certainly, if we pay attention to contemporary media reports about digital media, these reports shout out important, often frightening, ethical issues.

So, for example, my local newspaper recently reported on a local pastor – and mayor of a small town – who was captured in an online sting operation. An "Internet detective" posing as a young girl in a chat room enticed the 62-year-old man, hoping to meet up with the "girl" for sex, into revealing his identity (O'Dell 2008).

Somewhat more subtly, the Danish tabloid newspaper *Nyhedsavisen* (which enjoys one of the largest circulations in Denmark) recently reported that two new words have been added to the dictionaries of mobile phones – *fråderen* ("foaming," slang for "hungry") and *luder* (slut). The sub-headline to the story read: 'An expert wonders whether this means in the long run that young people's language will develop in a negative direction' (Mainz 2007: 14). (To be sure, an additional expert quoted in the story comments that this development is simply another reflection of how all languages change – and that such changes are not necessarily reasons for panic.)

What these examples illustrate is the tendency of popular media to call our attention to important ethical issues involved with digital media – but in ways that run the risk of fostering what are called "moral panics." That is, in order to attract our attention, such media stories focus unduly on the sensational (if not the sexual). But thereby, they tend to appeal to a deep-seated fear that our new technologies are somehow getting out of control (a fear that has been expressed in the modern West since Mary Shelley's *Frankenstein* ([1818] 1933) – in part, as these new technologies apparently threaten to corrupt our ethical and social sensibilities.

As we will see in subsequent chapters, this media approach is most pronounced (understandably enough) with regard to sex and violence – specifically, concerns about pornography online and violence in digital games (chapter 5). But here we can start to see how such a "moral panic" style of reporting both furthers and frustrates careful ethical reflection on digital media.

On the one hand, to be sure, such reporting succeeds in getting our attention – and thereby provides a useful service by catalyzing more careful reflection on important ethical issues evoked by digital media. On the other hand, by highlighting especially the potentially *negative* effects of digital media, such reporting fosters a polarized way of thinking – an approach that could be characterized as "technology good" (because it brings us important benefits) vs. "technology bad" (because it threatens the moral foundations of society, most especially the morality of young people). The problem is that such an approach to thinking about important ethical issues is simply misleading. As we will see – and as most of us probably already know full well – whatever truths may be discerned about the ethics of digital media often lie somewhere in the middle between these two extremes. But if we are only presented with the simple choice between "technology good" and "technology bad," we may be tempted to think that these are indeed our only choices and get stuck in trying to choose between two compelling alternatives. Getting stuck in this way short-circuits what we need to do if we are to move beyond such either/or thinking – a movement that requires more careful and extensive reflection.

One way to see how to move beyond such polarities is first to examine more carefully some of the important characteristics of digital media, along with the specific sorts of ethical issues that these characteristics often raise for us.

(Ethical) life in the digital age?

"Digital media" are the subject of an extensive range of analyses in a number of disciplines (e.g., Lievrouw and Livingstone 2006). This book, however, takes the standpoint of an interest in digital media ethics; thereby, at least at the beginning, we are interested in seeing the distinctive features of digital media – what sets them apart from earlier media – that make them ethically challenging and interesting.

To be sure, digital media represent strong *continuities* with earlier forms of communication and information media such as printed books, journals, and newspapers, what we now call "hard-copy" letters, and, for example, traditional forms of mass media that include not only newspapers but also "one-to-many" broadcast media such as radio and TV. We will note and explore these continuities more fully in our efforts to evaluate one of the larger *ethical* questions we will confront – namely, do digital media present us with radically *new* kinds of ethical problems that thereby require absolutely new ethical approaches? (See *Preliminary exercise*, below, p. 11.) For now, we can note that these questions are driven in good measure by rather emphasizing the important *differences* between earlier media and digital media. (Such an emphasis, by the way, also drives the "either/or" approach underlying much popular media reporting.) In particular, these differences often are part of why new ethical issues come up in conjunction with digital media. Exploring these differences at the outset thus seems like a good starting point.

Here, then, we will consider three distinguishing characteristics of digital media (though others are important): how digital media (in contrast with analogue media) foster *convergence;* digital information as "greased"; and digital media as (global) *communica-tion* media.

1. *Digital media, analogue media, and convergence*

To begin with, digital media work by transforming extant informa-tion (e.g., voices over a phone, texts written on a wordprocessor, pictures of an impressive landscape, videos recorded and broadcast, etc., etc.) into the basic informational elements of electronic com-puters and networks, including binary code (1's and 0's – bits on and off) and the definition of how such code is to be manipulated within a given application. Digital media contrast in this way with *analogue* media – such as an old-fashioned vinyl record – that capture, store, and make information accessible by producing specific artifacts that are *like* the original. In the case of music, recording equipment, beginning with microphones and conclud-ing with a storage medium such as audiotape, translates the vibrations of an original sound into magnetically stored informa-tion, corresponding to specific sound pitches and volumes, which is then "written" onto a tape that passes by a recording head at a specific speed. These *analogues* of an original sound are then in turn transformed into further analogues, as they are mechanically carved onto the grooves of a vinyl record in the form of bumps and valleys that correspond to (i.e., are analogues of) the high and low frequencies and volumes of the original sound. These physical vari-ations are further translated by a phonograph needle into electronic impulses that likewise mimic the original variations of a sound. Finally, these impulses are transformed into sound by an amplifier and speaker(s) – again, as an analogue or copy of the original that, if all goes well, is as close to the original as possible.

One way to think about analogue media is that they work by capturing, recording, and replaying information as a *smooth*, continuously variable content. That is, there are comparatively con-tinuous variations as a series of musical notes, first low and soft and then high and loud, are captured and replayed as moving evenly from the one to the other. In this sense, the contents of ana-logue media are more or less infinitely variable. By contrast, *digital* media capture, store, and make accessible their contents based entirely on a binary code of 1's and 0's.

More importantly, analogue media always involve some loss of information across the various processes of collecting, recording, and storing it. This means – and this is particularly critical to the ethical discussions of copying – that each analogue copy of an original is always less true to the original; and the more copies made – e.g., a tape copy of a record as a copy of a tape of an original performance – the less faithful (and satisfying) the resulting copy will be. By contrast, once information is transcribed into digital form, each copy of the digital original will be (more or less) a perfect replica of the original. Copy an MP3 version of your favorite song a thousand times, and if your equipment is working properly, there will be no difference between the first copy and the thousandth copy.

Even more importantly, analogue media are strongly distinct systems: how information is captured and replayed in a vinyl record is not immediately compatible with – and thereby, easily exchangeable with – how information is captured and replayed in a newspaper or printed book. By contrast, once information is translated into digital form, such information – whether destined for an MP3 player as an audio recording, or a wordprocessor as text – can be stored on and transmitted through a shared medium. Hence the same computer, PDA, or even smartphone, for example, can hold digital photos and music, along with wordprocessing files, spreadsheet files, etc.

As once distinct forms of information are thus translated into a commonly shared digital form, it makes possible one of the most important distinguishing characteristics of digital media, namely, convergence (e.g., Briggs and Burke 2005: ch. 7; Jenkins 2006; Storsul and Stuedahl 2007). Such convergence can be (literally) seen in a rich webpage that contains text, video, and audio sources, as well as possibilities for sending email, remotely posting a comment, etc. These once distinct forms of information and communication are now conjoined in digital form, so that they can be transmitted entirely in the form of 1's and 0's via the Internet. Similarly, a contemporary cellphone further exemplifies such convergence: it is capable of not simply handling phone conversations, but also handling digital information used for a built-in camera (still and/or moving video), MP3 player, web browser capable of capturing text and other sorts of information, etc.

This means, then, that digital media bring together both traditional and sometimes new sorts of information sources. In particular, what were once distinct kinds of information in the analogue world (e.g., photographs, texts, music) are now no longer strongly distinct; rather, they share the same basic *form* of information. What does this mean, finally, for ethics? Here's the key point: what were once distinct sets of ethical issues now likewise converge – sometimes creating new combinations of ethical challenges that we haven't had to face before.

For example, societies have developed relatively stable codes and laws for the issue of *consent* as to whether or not someone can be photographed in public. (In the U.S., generally, one can photograph people in public without asking for their consent, while in Norway, consent is required.) Transmitting that photo to a larger public – e.g., through a newspaper or a book – would then require a different information system, and one whose ethical and legal dimensions are addressed (however well or poorly) in copyright law. But as many people have experienced to their regret, a contemporary cellphone can not only record their status and actions, but further (more or less immediately) transmit the photographic record to a distribution medium such as a Facebook profile or even more public website. The *ethics* of both *consent* in photography and *copyright* in publication are now conjoined in new ways that we simply have not had to think through before.

Preliminary exercise: do we really need a (new) digital media ethics?

In response to these sorts of new ethical challenges brought about by digital media (and new technologies more generally), two sorts of arguments have emerged:

I. The new possibilities of photographing people in both public and more intimate situations, coupled with more or less immediately posting such photographs and/or videos to a forum such as a social networking site or more public webpage, means that people are now *more vulnerable* to violations of privacy.

That is, where privacy can be minimally defined as the capacity to control information about oneself, the new ability of others to record and quickly distribute potentially embarrassing information about oneself thereby decreases one's control over such information (e.g., in the form of permission to take a photograph, much less permission to distribute the photograph in a semi-public or public forum).

A general guideline in many ethical systems, as we will see, is that *increased vulnerability* requires *increased responsibility*. So, for example, parents have increased responsibilities for their children as infants that they do not have for their children as young adults, because children as infants are vulnerable in many ways that young adults are not.

On this line of thinking, the fact that others around me are *more vulnerable* in the presence of my cellphone argues that I as the owner/operator of the device need to be *more responsible* for how I use it. This might mean, for example, paying increased attention to whether or not what I'm recording and distributing might harm the person in some way; increased attention to the need for others to have and maintain control of information about themselves, etc.

More generally, this line of argument suggests that we now face a pressing need for careful and systematic ethical reflection, so as to develop the sorts of guidelines, codes, and laws that can help us work through especially the new sorts of ethical issues that digital media evoke.

2. Alternatively (but not necessarily to the exclusion of the first line of argument above), the history of technology is in part a history of people learning how to use new technologies in ethically appropriate ways – though often only after sometimes terrifically damaging (perhaps even fatal) experiences with them. On this view, we "muddle through" as a species, learning

more from our experience than from careful reflection on how to utilize new technologies in ways that minimize harm, protect rights, etc.

From this perspective, there is no real need to undertake ethical reflection on the new situations confronting us as new technologies enter our individual and social lives. Rather, we can generally rely on people's common sense and previous ethical experiences and judgment to help forge the new ethical guidelines that will become generally known and respected. For example, if I post an embarrassing photograph or video of a friend on the Web, thinking it's all in good fun, but thereby manage only to make my friend really angry and upset with me, I'll eventually figure out that I shouldn't do such things.

While, with the advantage of hindsight, I may well look back on such an act as a mistake – at least I'll avoid such errors in the future. There's no need for extensive ethical reflection ahead of time; rather, we will simply learn from our mistakes.

Questions for reflection/discussion/writing

1. Of the two positions outlined above, which better describes your approach to the ethical dilemmas that arise as you take up new technologies in your life?

   Provide an example of this if you can.

2. Do you have a reason or sense for why your approach might be better than the alternative? Explain your view here as fully as you can.

3. Can you think of an instance or example – whether in relation to digital media technologies or to other new technologies – where your approach might *not* work as well as the alternative?

4. Do you have other thoughts and suggestions for how we – both individually and as a society and species – might best approach the ethical dimensions of digital media (and/or other new technologies)?

2. Digital media and "greased information"

A second characteristic of digital media is that the information they capture, record, and transmit is "greased." That is, as James Moor has observed, "When information is computerized, it is *greased* to slide easily and quickly to many ports of call" (Moor 1997: 27). As anyone who has hit the "send" button on an email too quickly knows all too well, information in digital form can spread more or less instantaneously *and* globally, whether we always want it to or not.

As the example of uploading potentially embarrassing photos or videos from a cellphone suggests, the capacity of digital information to move quickly from one place to another raises especially serious ethical issues surrounding *privacy*. That is, where it was once comparatively difficult to capture and then transmit information about a person that s/he might consider private, the advent of digital media, beginning with computer systems that can store and make easily accessible a wide range of information about persons, has resulted in a wide range of new threats to what was once clearly personal and private information. Moreover, digital information as "greased" likewise makes it easy to copy and distribute, say, one's favorite songs, movies, or texts. To be sure, it has always been possible to copy and distribute copies of a given text – or, in the days of analogue media, of a given song or film. But the *ease* with which we can do so by way of digital media appears to be one factor that makes such copying and distribution an even more pressing ethical problem these days.

(We will examine the issues of privacy more fully in chapter 2, and the issues of copyright in chapter 3.)

3. Digital media as communication media: global scope and interactivity

The emergence of digital media – along with the Internet and the Web as ways of quickly transporting digitized information – thus gives rise to strikingly new ways of communicating with one another at every level. Emails, social networking sites (Facebook,

MySpace, etc.), photo and video distribution sites (Flickr, YouTube, etc.), and personal blogs provide ways for people – especially in the developed world but also increasingly in developing countries – to substantially enhance existing relationships and to develop new ones with persons often far removed from their own geographical/cultural/linguistic communities. Especially as the Internet and the Web now connect more than one-sixth of the world's population, they thereby make possible cross-cultural encounters online at a scope, speed, and scale unimaginable even just a few decades ago.

Along these lines, two additional features of digital media become crucial. To begin with, digital media enjoy what Phil Mullins (1996) has characterized as a kind of fluidity: specifically, a biblical text in digital form – either on one's PDA or as stored on a website – becomes, in his phrase, "the fluid Word." In contrast with a biblical text as *fixed* in a strong way when inscribed on parchment (the Torah) and/or printed on paper, a biblical text encoded on a flash memory or server hard drive in the form of 1's and 0's can be changed quickly and easily. This fluidity is highlighted by a second characteristic of digital communication media – namely, interactivity. Both a printed Bible and the daily newspaper are produced and distributed along the lines of a "top-down" and "one-to-many" broadcast model. While readers may have their own responses and ideas, they can (largely) do nothing to change the printed texts they encounter. By contrast, I can change the biblical text on my PDA if I care to (e.g., if I think a different translation of a specific word or phrase might be more precise or illuminating) – and, by the same token, a community of readers can easily amend and modify an online text; they might also be able to post comments and respond to a given text in other ways that are in turn then "broadcast" back out to others. To put it differently, digital communication media provide us with new possibilities of "talking back": posting comments, or even a blog, in response to a newspaper story, now reproduced online; voting for a favorite in a TV-broadcast contest by way of SMS messaging; organizing "smart mobs" via the Internet and cellphones (as well as older media such as fax machines) to protest – and, in some cases, successfully depose – corrupt politicians, etc.

Especially as fluid and interactive digital media simultaneously enjoy a global scope, they evoke a host of important ethical issues: because our communications can quickly and easily reach very large numbers of people around the globe, our use of digital communication technologies thus makes us *cosmopolitans* (citizens of the world) in striking new ways. In this new context, we are forced to take into account the various and often very diverse *cultural* perspectives on the ethical issues that emerge in our use of digital media. So I will stress throughout this book how the assumptions and ethical norms of different cultures shape specific ways of reflecting on such matters as privacy (chapter 2), copyright (chapter 3), and pornography and violence (chapter 5). In addition, in chapter 4, we will will look specifically at the ethics of cross-cultural communication online, beginning with the example of the so-called "Muhammad cartoons," published by a relatively small-circulation Danish newspaper, *Jyllands-Posten,* in 2006. Had these cartoons been restricted to print media, they would have likely had little impact outside of Denmark. But as made immediately available online, they sparked world-wide diplomatic furor and violent protest, resulting in dozens of deaths and considerable property damage. This example makes very clear how a globally distributed digital media thus amplifies the consequences of our communicative acts far beyond the boundaries familiar to us with traditional media. Again, digital media thus bring before us new considerations that are ethically relevant to familiar ethical issues – and perhaps confront us with distinctively new sorts of ethical issues as such.

Digital media ethics: how to proceed?

At first glance, it might seem that developing such an ethics is an impossible task. For one thing, especially because digital media present us with often strikingly new sorts of interactions with one another, it is not at all clear whether – and if so, then how – the ethical guidelines and approaches that are already in place (and comparatively well-established) for traditional media would apply.

In addition, digital media as global media thus force us to confront culturally variable views – not simply regarding basic ethical norms and practices, but, more fundamentally, regarding how ethics is to be done. In particular, we will see that non-Western views – represented in this volume by Confucian, Buddhist, and African perspectives – challenge traditional Western notions of the primary importance of the *individual*, and thereby Western understandings of *ethical responsibility* as primarily *individual* responsibility. That is, while we in the West recognize that multiple factors can come into play in influencing an individual's decision – e.g., to tell the truth in the face of strong pressures to lie, to violate another's rights in some way, etc. – we generally hold *individuals* responsible for their actions, as the *individual agent* who both makes decisions and acts independently of others. But these days, our interactions with one another increasingly take place via digital media and networks. This means, more specifically, that multiple actors and agents – not only multiple humans (including software designers as well as users), but also multiple computers, networks, 'bots, etc. – must work together to make specific acts (both beneficent and harmful) possible. So it appears that, in parallel with the distribution of information via networks, our *ethical responsibility* may be more accurately understood in terms of a *distributed responsibility.* That is, in contrast with a single person taking all the responsibility for his or her actions, ethical responsibility for the various actions we are able to undertake via digital media and networks may be shared, so to speak, across the network. We will see that this understanding of distributed responsibility is, in fact, not a new idea; rather, it is one shared with both pre-modern Western philosophies and religions and multiple philosophies and religions around the globe.

On the one hand, this is a Very Good Thing, as it may point towards important ethical norms and practices that can be shared among the multiple cultures and peoples now brought into communication with one another through the Internet and the Web. But, on the other hand, it represents a major challenge especially to Western thinkers used to understanding ethical responsibility in primarily individualistic terms.

Is digital media ethics *possible?* Grounds for hope

These challenges are certainly daunting; indeed, when we first begin to grapple with digital media ethics, especially with a view towards incorporating a range of global perspectives, the tasks before us may seem to be overwhelming and, we may be tempted to think, futile.

But both our collective experience with earlier technological developments and more recent experience in the domain of Information and Computing Ethics (ICE) suggest, despite the considerable challenges of developing new ethical frameworks for new technologies, we are nonetheless able to do so. Indeed, this experience provides us with a number of examples of ethical resolutions that "work" both globally (as they involve discerning shared norms and understandings) and locally (as they further involve developing ways of interpreting and applying shared norms in specific cultural contexts – and thereby preserving the distinctive ethical differences that define diverse cultural identities).

Two examples may be useful here. To begin with, as the Internet and Web first emerged in the 1990s, more and more researchers from a range of disciplines and cultures/countries sought to study the rapidly expanding ways in which humans interacted with one another online. Internet research thus developed and expanded – leading, for example, to the founding of the international Association of Internet Researchers (AoIR) in 1999. Along the way, however, more and more researchers were forced to wrestle with often new ethical issues concerning, for example, whether or not they were obliged to protect the identity and privacy of their subjects, whether or not their subjects enjoyed the right to informed consent, and so forth. These ethical demands were familiar to researchers in medicine and the social sciences, but as they were developed and applied to embodied subjects in real-world experiments and research. Some argued that insofar as researchers only examined what happened online in so-called "virtual" environments and communities, no "real" human subjects were involved – and hence the ethics of traditional human subjects protections did not apply. Others argued – supported by growing evidence that, in fact, our online behaviors and expressions are closely tied to our offline lives and selves – that some version of human subjects' protections should apply to online research as well.

In 2000, the AoIR began a two-year project to develop ethical guidelines for Internet researchers. Briefly, in this project we (I was chair of the committee, 2000–5) faced the range of difficulties that confront similar efforts to develop ethics for digital media, beginning with the need to determine how far traditional ethical frameworks may – and may not – successfully resolve the issues evoked by digital media and their new possibilities for communication, human interaction, and so forth. In addition, research on Internet-based phenomena is undertaken by researchers from around the globe; this means that when these researchers encounter ethical difficulties, they thus bring to bear on their ethical reflections a wide range of culturally variable ethical traditions and frameworks, along with often diverse laws (e.g., concerning privacy and data privacy protection).

In the face of such disciplinary and cultural diversity, how on earth could anyone imagine developing a single set of ethical guidelines that could be endorsed by researchers from around the globe?

Within two years, however, an interdisciplinary and international committee (representing 11 countries both East and West) developed a set of ethical guidelines for Internet research that were in fact endorsed by the AoIR membership (AoIR Ethics Working Group 2002). The AoIR guidelines have subsequently found significant use by students and researchers across a range of countries and cultures, suggesting that they are recognized as useful in resolving a number of ethical issues evoked in Internet research, even though these may arise in widely diverse cultural and national domains.

More recently, an emerging global dialogue among philosophers and ethicists regarding privacy and data privacy protection rights has highlighted, as we might expect, considerable differences between diverse cultures regarding our understandings and expectations surrounding individual and collective privacy. But as we will see in more detail in chapter 2, this global dialogue appears to be converging on a set of shared norms and agreements regarding the nature of privacy. Importantly, these shared norms and agreements are understood, interpreted, and applied in different ways in different cultures and nations. That is, privacy in Thailand, China, and Japan is understood in ways that reflect the basic cultural assumptions and national norms of these countries – and thereby preserve the distinctive identities and traditions of these cultures. Again, these understandings of privacy differ considerably from their counterparts in Western countries, such as the United States and the European Union – where, as we would expect, prevailing understandings of privacy reflect the underlying cultural assumptions and norms of these nations. Nonetheless, insofar as these Eastern and Western understandings derive from *shared* norms and conceptions of privacy, the result is an *ethical pluralism* that thereby conjoins shared ethical norms alongside important (indeed, defining) cultural differences. (See chapter 2 for further discussion of privacy; and chapter 6 for further discussion of ethical pluralism.)

How to do ethics in the new mediascape: Dialogical approaches, difference, and pluralism

These examples of the AoIR guidelines and an emerging pluralism regarding privacy suggest that we can undertake the enterprise of digital media ethics with some hope of success. As well, I think that we can further draw from these examples important suggestions for how to proceed. Both examples, in fact, share two elements in common. To begin with, they each incorporated what we can think of as *dialogical* approaches – approaches that emphasize the importance of listening for and respecting *differences* between our diverse ethical views.

Ordinarily – especially if our thinking is shaped by a polarized "either/or" common in popular media reporting – we tend to understand the *difference* between two views in only one possible way: if the two views are *different*, one must be right and the other wrong. As we will explore more carefully in chapter 6, such an approach is called ethical *absolutism* or ethical *monism*. Such absolutism or monism may work well in certain contexts and with regard to some ethical matters. But especially in a global context, a frequent, limiting consequence of such ethical monism is to force us into thinking that one and only one particular ethical framework and set of norms and values (usually, those of the culture(s) in which we grew up) are *right*, and those that are *different* can only be *wrong*.

Oftentimes, in the face of such monism and its intolerance of different views, we are tempted to take a second position – one called ethical *relativism*. Ethical relativism argues that beliefs, norms, practices, frameworks, etc., are legitimate in relation to a specific culture; in this way, ethical relativism allows us to avoid the intolerance of ethical monism, and to accept all views as legitimate. Such an approach is especially attractive as it prevents us from having to judge among diverse views and cultures: we can endorse all of them as legitimate, in at least a relative way (i.e., relative to a specific culture, time, and/or place).

But the examples we have seen of ethical *pluralism* in both Internet research ethics and emerging understandings of privacy and data privacy protection make clear that such pluralism stands as a third possibility – one that is something of a middle ground between absolutism and relativism. That is, to begin with, such pluralism avoids the either/or of ethical monism – an either/or that forces us to choose between two different views, endorsing one as right and the other as wrong. Rather, it is possible to see that different views may emerge as diverse interpretations or applications of *shared* norms, beliefs, practices, etc. Insofar as we can discern that this is so, the differences between two (or more) views thus do *not* force us to accept only one view as right and all the others as wrong. Rather, we can thereby see that many (but not necessarily all) different views may be right, insofar as they function as diverse interpretations and applications of shared norms and values.

In addition (though this may be a little confusing at this stage), ethical pluralism thereby overcomes a second "either/or" – namely, the apparent polarity between ethical *monism* and ethical *relativism* themselves. That is, when we first encounter these two positions – and, again, especially if our thinking has been shaped by prevailing dualities in the thinking of those around us, including popular media reports – our initial response may again be either/or: either *monism* is right or *relativism* is right, but not both. In important ways, ethical pluralism says that both are right, and both are wrong. From a pluralist perspective, monism is correct in its presuming that universally valid norms exist, but mistaken in its insistence that the differences we observe between diverse cultures in terms of their practices and behaviors must mean that only one is right, and the rest wrong. Similarly, from a pluralist perspective, ethical relativism is correct in its attempt to endorse a wide range of different cultural norms and practices as legitimate, but mistaken, first of all, in its denial of universally valid norms.

Again, we will explore these theories of absolutism, relativism, and pluralism in much more detail in chapter 6. Here it suffices simply to introduce these possibilities of thinking in an initial way as we seek to move beyond the either/or thinking that tends to prevail in popular media and thereby tends to shape our own thinking.

Given this first introduction, perhaps we can now see more clearly why the "either/or" underlying much of popular media report works *against* our best thinking. Ethical pluralism requires us to think in a "both/and" sort of way, as it conjoins both shared norms and their diverse interpretations and applications in different cultures, times, and places. But if the only way we are able to think about ethical matters is in terms of the "either/or" of ethical monism, then we literally cannot conceive of how to move beyond the right/wrong dualisms it often confronts us with. That is, we will find it difficult conceptually to move towards pluralism and other forms of middle grounds, because our either/or thinking insists that we can only have either unity (shared norms) or difference (in interpretation/application), but not both.

Stated differently: in dialogical processes we emphasize learning to listen for and accept *differences* – rather than reject them from the outset because different views must thereby be wrong (ethical monism). But neither do we come to endorse all possible views as correct (ethical relativism), because not every view can be understood as a legitimate interpretation or application of a shared norm. Rather, dialogical processes help us sort through which views may stand as diverse interpretations of shared norms in a pluralism, and those views (e.g., endorsing genocide, racism, violence against women as inferior, etc.) that cannot be justified as interpretations of shared norms.

Doing ethics: further considerations

Another difficulty with the "moral panics" approach to ethical issues in the new mediascape is that it suggests that "ethics" works like this:

1. There are clear, universally valid norms of right and wrong that we can take as our ethical starting points – as *premises* in an ethical *argument.*[2]
2. All that "ethics" really involves is applying these initial premises to the particulars of the current case in front of us – in a straightforward deduction that concludes the right thing to do, as based on our first premises.
3. Once we have our ethical answers in this way, we can be confident that our answers are right; those who disagree with us must be wrong.

This approach to ethics, I would suggest, is not necessarily mistaken; on the contrary, it seems to me that much of the time, most of us in fact do not perceive an ethical problem or difficulty in the situation we're facing – because our ethical frameworks already

---

2 Here I use the terms "premise," "argument," "conclusion," etc., in their logical sense. An understanding of the basic element of logic is essential for undertaking ethics – and many ethics texts include an introduction to logic (e.g., Boss 2005; Tavani 2007: ch. 3, etc.). For the sake of brevity I have chosen instead to introduce and discuss a minimal number of logical elements: *analogy* and *questionable analogy* in chapter 3; the distinction between *exclusive* and *inclusive* 'or's in chapter 5; and the basic fallacy of *affirming the consequent* in chapter 6. Otherwise, in addition to any preferred resources of instructors, I would further recommend some of the excellent introductions to logic that already exist (e.g., Weston 2000; Bowell and Kemp 2005; Possin 2005).

provide us with reasonably clear and straightforward answers along just these lines. Most of us, for example, do not routinely lie, steal, or kill – despite sometimes what may be considerable temptations to do so – because we accept the general norms and principles that forbid such acts.

At the same time, however, this initial understanding of ethics obscures a number of important dimensions of ethical reflection. Among other things, this initial approach runs counter to what seems to actually happen when we encounter genuine ethical problems and puzzles. Take, for example, the notorious problem of downloading music illegally from the Internet. More or less everybody knows that this is illegal, but we are also pulled and influenced in our thinking by other considerations, e.g.:

> I'm not likely to get caught, so there's virtually no possibility that this will actually hurt me in some way.

> The internationally famous musicians – and the multinational companies that sell their music as product for profit – are certainly wealthy enough. They won't feel the loss of the 2 cents profit they would otherwise enjoy if I paid for the music.

> Copyright laws are unfair in principle: they are written for the advantage of the big and already wealthy countries. Thus I think illegal downloading by a struggling student in a developing country is a justified form of protest against multinational capitalism and its exploitation of the poor.

> Whatever the law says, the law is the law: I think it should be respected so far as possible – not only in order to avoid punishment, but in order to thereby contribute to good social order.

> Even if the chances of getting caught are vanishingly small, if I *do* get caught, the negative consequences would be enormous (fines, possibly problems at work, maybe even jail time). It's not worth breaking the law to save a few bucks on music.

> While internationally famous artists may not miss my contribution to their royalties, local and/or new artists certainly will. I'll not rip them off by illegally copying their music – I'll just order the song online or buy the CD instead.

The point here is not only that we are often pulled in competing directions by values and principles that appear to contradict one another; the more fundamental problem is:

> given the specific details of our particular situations, how do
> we know *which* principle, value, norm, rule, etc., is in fact
> relevant to our decision?

This is to say, in direct contrast with the "top-down" *deductive*
model of ethical reasoning – i.e., one that moves from given
general principles to the specifics of our particular case – this
second ethical experience begins with the specifics of our particular case, in order to then try to determine ("bottom-up") which
general principles, values, norms, etc., in fact apply.

This second maneuver is thereby far more difficult, as it first
requires us to *judge* – based on the particulars of our case – which
general principles, norms, values, etc., apply to our case. Clearly,
without such general principles, we cannot make a reasoned
decision. But the great difficulty is this:

> there is no general rule/procedure/algorithm for discerning
> which values, principles, norms, approaches *apply*; rather,
> these must be discerned and *judged* to be relevant in the
> first place, *before* we can proceed to any inferences/
> conclusions about what to do.

Aristotle referred to the kind of judgment that comes into play
here as *phronesis* – a term often translated as "practical judgment."
For Aristotle (and for many ethicists in multiple traditions around
the world), the development of this sort of practical judgment –
i.e., one that can help us discern in the first place just which norms
and values do apply to the particulars of a specific case – is an on-
going project that goes on throughout one's whole life. This is in
part because it requires *experience* – both of successes and failures
– as these help us learn (oftentimes, the hard way) what "works"
(is relevant) ethically and what doesn't. The first time we try to
learn a new skill or ability – say, ice-skating – we are certain to
stumble and fail, perhaps catastrophically, and almost certainly
more than once. Analogously, our first efforts to grapple with
difficult ethical issues that require *phronesis* do not always go well:
we are caught in the ethical "bootstrapping" problem of needing
precisely the ability to judge that is robust enough to help only

after it has been developed and honed through many years of (sometimes hard) experience.

But as we also say: we have to start somewhere . . .

Overview of the book, including suggestions for use

By now, I hope readers have a reasonably clear idea of the features of digital media that lead to specific sorts of ethical issues that we will explore more fully in subsequent chapters. I also hope that you are beginning to have a sense that as we do so, especially with regard to digital media that interconnect us globally, it is important to do so in ways that go beyond the either/or polarities that tend to dominate popular media reporting.

At this stage, it may also be helpful to understand why the chapters of the book are arranged as they are. The book is organized in a some-what unusual way, but one that I hope will be more effective and useful. I follow here what I think of as a "circle" approach to explor-ing and teaching ethics, one that intentionally moves back and forth between (a) specific, real-world examples from how we actually use digital media and thereby encounter specific ethical problems and (on a good day) legitimate resolutions, and (b) a number of relevant *theories* precisely about how we are to resolve such ethical challenges and difficulties. Ethics texts often begin with a rather complete listing and discussion of the important theories, on the very sensible view that students can come to grips with specific ethical difficulties only if they are first familiar with a complete range of ethical theo-ries. Here, I've done the reverse, and placed the resources on ethical theory at the end of the text (chapter 6). The idea is to encourage stu-dents and instructors to take up just two or three of these theories at the beginning, and then apply them to the specific cases taken up in each of the chapters. After students are thereby comfortable with how two or three theories work in their application to real-world cases, they can then return with their instructor to take up additional theories – and then apply these theories to additional cases. While I offer a specific reading plan in this direction (below), by placing the theory/meta-theory chapter at the end of the text, I hope to provide students and their instructors with more flexibility in determining for themselves as they go along just how much theory they wish to absorb vis-à-vis specific issues and problems.

This organization reflects my own teaching experience. I have learned (over several decades) that my students are more likely to learn both an extensive range of theories and how to apply them if we rather begin with just two or three theoretical frameworks, and then apply these to specific cases. To be sure, our grappling with specific cases early in the course will thus be hampered in the sense that my students will not have a more complete range of theory at their disposal. But at the same time, it often happens that they will discover that the theories they initially bring to bear on a given case are *not* adequate – i.e., the initial theories do not allow them to resolve the problems in ways that better fit their own ethical intuitions and sensibilities. In short: from here, they see on their own the need for further theory/theories, and so as we return from specific cases to more theories (making the circle from *praxis* to theory), they are characteristically more interested in and open to learning about new theories than they would be if we simply worked through all of them from the outset.

By the same token, nothing prevents us from going back to re-consider earlier cases in light of more recently acquired theories – and thereby seeing these cases in a new light (making the circle from theory to *praxis*). Indeed, when things work well, our doing so helps us come to resolutions of the ethical problems involved in more satisfying ways than previously – thereby enhancing our appreciation not only for how a specific theory may offer distinctive advantages vis-à-vis a specific case, but also for how a now greater range of theories *work* in their application to real-world issues and problems.

Suggested reading plans

Circle method: Instructors and their students who want to follow this approach can do so by moving into chapter 2 (privacy), before turning to chapter 6 for a first run at some of the most important theories. For both chapters 2 and 3 (copyright and intellectual property), you will find the introductions to *utilitarianism* and *deontology* to be the minimum needed from chapter 6; but you may, depending on your interests and contexts, find other(s) to be of importance and interest in this first foray as well. It's also the case that *ethical pluralism* is central to chapter 2, and so it will be helpful to take up the section on ethical pluralism in chapter 6 in conjunction with chapter 2.

From there, you can return to chapter 3 – and/or turn further in chapter 6 for more theory, meta-theory, and especially the material on non-Western ethical traditions (Buddhism, Confucian thought, and African thought). This much from chapter 6 should be completed prior to chapter 4 (the ethics of cross-cultural communication online). As well, the material on *virtue ethics* from chapter 6 is presupposed for chapters 4 and 5 (pornography and violence). But some readers may prefer to go to chapter 5 prior to chapter 4 (or 3, for that matter), as more concrete and specific in certain ways, before taking up chapter 4 (or 3).

"Traditional" method: Instructors and students who prefer to plunge first into matters of ethical theory and meta-theory should simply start with chapter 6 first, before then turning to any of the specific chapters (2–5).

Case-studies; discussion/reflection/writing/research questions

Each chapter includes real-world examples intended to elicit initial reflection; and so these are accompanied by often an extensive series of questions and suggestions for "reflection/discussion/writing/research." I hope that these questions and suggestions can be used in an initial way by students and classes as simply catalysts for reflection, discussion, and perhaps informal writing. Instructors may also find useful suggestions here for questions and material that can be developed into more formal writing and research assignments. But these are only "starters" and examples. Instructors and students will certainly come up with their own preferred questions, case-studies, etc. (And if you do, please share!)

In general, I suggest that students have a chance to reflect and write through their initial responses to opening case-studies and examples prior to further reading into the chapter. The intention is to help students (and their instructors) articulate their initial intuitions and sensibilities as a starting point for discussion and further reflection. Of course, students (and their instructors) may eventually change their views, but our initial intuitions and views are where we must start, and it helps to articulate these as fully and clearly as possible before moving on.

O.k. – enjoy!

CHAPTER 2

Privacy in the Electronic Global Metropolis?

Everyone has the right to respect for his private and family life, his home and his correspondence.

(Council of Europe, European Convention for the
Protection of Human Rights and Fundamental Freedoms,
Section I, Article 8 [1950])

In a democracy, privacy is a basic political right that cannot be sold out in the marketplace.

(Reidenberg 2000)

Under ubuntu *[an African worldview emphasizing connectedness and community welfare over individual welfare], personal identity is dependent upon and defined by the community. Within the group or community, personal information is common to the group, and attempts to withhold or sequester personal information are viewed as abnormal or deviant. While the boundary between groups may be less permeable to information transfer,* ubuntu *lacks any emphasis on individual privacy.*

(Burk 2007: 103)

The freedom and privacy of correspondence of citizens of the People's Republic of China are protected by law. No organization or individual may, on any ground, infringe upon the freedom and privacy of citizens' correspondence except in cases where, to meet the needs of state security or of investigation into criminal offences, public security or procuratorial organs are permitted to censor correspondence in accordance with procedures prescribed by law.

(Constitution of the People's Republic of China 1982,
Article 40, cited in Lü 2005: 9)

You have no privacy, get over it.

(Scott McNealy, co-founder, Sun Microsystems,
cited in Springer 1999)

Chapter overview

This sampling of diverse perspectives on privacy leads us into initial reflections and then exercises on privacy and anonymity online. Following some cautions regarding the notion and possible (mis)uses of "culture," we then explore how different people in different cultures understand and value privacy in different ways – including important differences between Eastern and Western conceptions, as well as between the United States and the European Union. In this context, finally, we first explore the important meta-ethical positions of ethical absolutism, ethical relativism, and ethical pluralism. We see how these positions shape our responses to the *diversity* of cultural views and beliefs regarding privacy – diversity that must be preserved, in my view, alongside any effort to establish a *global* ethics of norms and practices that are shared around the world.

Information and privacy in the global digital age

1. Privacy and anonymity online – is there any?

Many people assume that their email communications are more or less private. We all know, of course, that someone with specialized software and hardware tools (such as a hacker and/or a government agency "sniffing" for information) may look into your email. Apart from these sorts of positive actions, though, we might assume that we can trust that our email is sent and received as *private* information.

In addition, many users may believe that they can send and receive information via email with their identity protected. Indeed, many of us know how easy it is to set up a free email account with a popular service such as Gmail: we can choose any user name we like, for example – including simple *pseudonyms* that appear to hide our real identity. Such pseudonymity or anonymity might be very helpful – especially for political purposes. Democratic activists in countries with authoritarian regimes might be able to use email to exchange ideas and information considered by the authorities to be dangerous, to help organize protests and other important actions, etc. With their real identities hidden behind a pseudonym, they could do so with greater freedom and safety. (This sort of scenario, indeed, is part of the original intuitions as to why *privacy* is a crucial right to be protected in democratic societies.)

But many users do not seem to be aware, for example, that their email contains a great deal of information about them – including information that can be used to determine their identity – and that such information is essentially public.

For example, most email clients (i.e., the software packages such as Outlook, Outlook Express, Entourage, [Apple] Mail, and Thunderbird) are set to show users only the basic contents of an email: sender, recipient, cc's, subject line, and email body. But look again: these clients also allow you to review the complete contents of your email – usually by selecting, e.g., "Source" under the view menu in Thunderbird or Entourage, once a given message is selected. (Outlook Express requires more steps: once a message is selected, look under the "File" menu for "Properties" and select this option. Then click on the "Details" tab.)

Initial reflection/discussion/writing

Select a recent email from a friend, and then view the complete source of the email as outlined above.

[*Nota bene*: if you use a webmail application such as Gmail, you may not be able to see the complete source in any easy or straightforward way. As we are about to see, there is good reason for this.]

A. What strikes you about the information contained here in the lines prior to the usual information about sender and receiver addresses?

B. Notice that each email includes here a history of how it was sent, usually in a format like this (from a recent email using a free guest account):

```
Received: from n9a.bullet.ukl.------.com
(n9a.bullet.ukl.------.com [217.146.183.157])
  by gulch.more.net (Postfix) with SMTP id CFA87C3CC7
  for <cmess@drury.edu>; Sun, 20 Jan 2008 14:21:53 -
0600 (CST)
Received: from [217.12.4.215] by n9.bullet.ukl.-----
.com with NNFMP; 20 Jan 2008 20:15:24 -0000
Received: from [216.252.122.217] by t2.bullet.ukl.---
--.com with NNFMP; 20 Jan 2008 20:15:24 -0000
Received: from [69.147.65.165] by t2.bullet.sp1.-----
.com with NNFMP; 20 Jan 2008 20:15:23 -0000
```

In particular, notice the IP (Internet Protocol) address: 69.147.65.165. What, if anything, does this IP address tell you?

[*Nota bene*: if you are *not* able to easily find this information using your mail client and/or in a specific email – i.e., depending on how your friend's email service works – again, we are about to see there may be a very good reason for this.]

Additional information, discussion

You may already know that such an IP address is essential to each and every transmission of information – whether an email, the text of a chat message, video, music, etc. – that takes place via the Internet.

You may further know that your computer is assigned an IP address by your Internet Service Provider (ISP) – usually a "dynamic" address, one that changes in a small way each time you log in to the ISP. Sometimes, the address is "static" – one permanently assigned to your particular machine.

But if you've ever wondered why, when you're browsing a website, you receive ads (either within a page and/or as a pop-up) that seem to "know" your physical location, the simple fact is: they do. That is, as your machine exchanges information with a website through your browser, your IP address is sent along as well. Even if you have a dynamic address, this change still belongs within a set of addresses assigned to a given ISP in a given area. "*They*" *know where you are* (at least approximately. . .).

It is for this reason, then, that Gmail, for example, justifies *not* revealing users' IP addresses in mail headers:

Personal information, including someone's exact location, can be gathered from someone's IP address, so Gmail doesn't reveal this information in outgoing mail headers. This prevents recipients from being able to track our users, or uncover what may be potentially sensitive personal information. ("Why don't you reveal users' IP addresses?", <http://mail. google.com/support/bin/answer.py?answer=26903&topic=1278>, accessed September 10, 2008)

By contrast, the service I used to create the example email header with IP addresses above clearly does reveal this information.

To be sure, if your IP address is assigned to your machine dynamically, then it is not a straightforward matter to tie your current IP address with your physical identity. But this leads to an important difference in how diverse countries and cultures decide to treat IP addresses. Currently, U.S.-based Google argues that IP addresses are *not* personal information and hence do not fall under the Data Privacy Protection codes of the European Union. But the European Union's group of data protection commissioners have recently concluded that IP addresses *are* personal information and thus require data privacy protection (White 2008a).

We will see more fully below that there are indeed strong *cultural* influences on how we understand and value "privacy" – including strong differences between the United States and the European Union, as reflected in and suggested by this example. However these contrasting views may be resolved (if at all), here we can continue by noting that it is possible to establish and sustain at least some level of anonymity online, for example, by using "anonymizer" software and webmail and web-browsing services that hide this sort of information. As well, by using encryption software (e.g., "Pretty Good Privacy" software, which is freely available), users can send emails that can be easily deciphered and read only by recipients who have been given the required encryption key, thereby assuring themselves a reasonably strong degree of privacy. But unless users take these unusual – and, in my experience, not widely familiar – steps, their transmissions across the Internet will thus be more or less open to anyone who cares to look.

Similar comments hold, by the way, for people using their computers to share information through peer-to-peer (p2p) networks –

such as those used, for example, in downloading and uploading music and other files through a network such as Gnutella or BitTorrent, but also in instant messaging exchanges, including the use of video cameras for video chat and conferencing.

Reflection/discussion/writing

C. Given what you now know about the information that you send along when you send an email or browse a website, how *private* do your interactions online seem to be?

In particular, compare the sorts of privacy available to you online with the privacy associated, for example, with a letter in a sealed envelope sent through governmental or private delivery services.

Does it seem to you that you enjoy *enough* privacy to under-take the sorts of activities you wish – and/or, more strongly, that you believe you have a *right* – to undertake, including what you do online?

And/or: does it seem to you that more privacy online might be a good thing – i.e., that it might be *ethically* required and justified?

possibly, but I'm not sur its necessa b/c if you

D. Whatever your responses in (C), explain *why*. That is, what *arguments/evidence/reasons* and/or other grounds, including feelings or intuitions, can you appeal to that would justify your response(s)? *– for the most part its private + have nothing the adds are only the to hide? helpful.*

Additional experience

Look up the privacy policies of the email service you use – whether it's a university-related service, and/or one of the many "free" services such as Gmail, Yahoo, etc.

Beyond what these policies say regarding *what* information about you is collected and shared, be sure to notice the *justifica-tions* for various forms of information-sharing.

(In the case of the service I used to generate the example email above, these justifications include the need to be able to: to diagnose technical problems; to send advertising messages to you that are targeted to your specific geographical region; to collect (aggregate) information to report to and/or sell to advertisers; and to control access to specific kinds of content.) ✓ all fine no prob

Reflection/discussion/writing

E. Any surprises here?
    In particular,

(i) is there either a kind of information about you and/or
(ii) a use of that information allowed for under the privacy policy that is news to you – i.e., that you were not aware of?

Identify these clearly.

F. Insofar as the policy you examine offers an ethical *justification* for its practices, highlight at least one of these.
    Either individually and/or as a small group/class, analyze the argument(s) at work here, especially with a view towards

(i) discerning what *kind* of ethical argument is offered (e.g., *utilitarian, deontological*, etc.); and
(ii) determining whether or not you (and your cohorts) find the argument persuasive.

If so, explain why. If not, explain why not.

2. Privacy: a matter of culture?

As the example of Google vs. the E.U. regarding IP addresses suggests – and as we might expect – our understandings of privacy vary widely, not simply from individual to individual, but also from culture to culture. The following exercise is intended to give you and your cohorts an initial set of indicators of where your sensibilities regarding privacy might lie upon a continuum of possibilities.

Consider the "smart ID" project of Thailand – a project that aims to create and issue national identity cards that contain the following information:

Name _____
Address _____
Date of Birth _____
Religion _____
Blood Group _____
Marital Status _____
Social Security _____
Health Insurance _____
Driver's License _____
Taxation Data [income bracket, taxes paid/owed] _____
Healthcare Entitlements _____
Officially Registered Poor Person? _____

(Kitiyadisai 2005: 32)

Reflection/discussion/writing

A. *Where do you draw the line?* Beginning with your own responses, which of these elements of identity would you be comfortable having encoded on a chip in a national ID card?

   Which of these elements do you think/feel/believe should *not* be included in a national ID card?

B. *Why?* In both cases, what *arguments/evidence/reasons* and/or other grounds, including feelings or intuitions, can you appeal to that would justify your response(s)? tax, Soc Sec, poor

C. You may want to compare your and your cohorts' sensibilities with the following that I've observed in using this example with students and faculty from a variety of cultural backgrounds.

   Roughly, reactions/responses range from a *minimum* to a *maximum* amount of information designated as public or private. These variations, moreover, appear to correlate with a number of values and sensibilities that are known to vary from culture to culture. One of the most important is suggested in the opening set of quotes, in the contrast between the Council of Europe's articulation of what amounts to *individual* privacy as a human right, vis-à-vis a lack of emphasis on individual privacy in the worldview of *ubuntu*, for example. Indeed, we will see that this lack of emphasis on individual privacy – in part, because of a greater emphasis on community harmony and integration – is characteristic of a wide range of non-Western cultures and traditions.

And within the domain of Western countries and cultures, there are further variations in our expectations regarding privacy that correlate with often very different understandings of the role of the state vis-à-vis the life of the individual.

So, for example, most *U.S. students* – if they accept the idea of a national identity card at all – are moderately comfortable with a card that would contain name, address, date of birth, and social security number. Perhaps religion. Perhaps blood group (in case of a medical emergency). Perhaps driver's license. Perhaps marital status. But it becomes unclear how much the federal government – or anyone else, for that matter, besides the person who handles my medical bills – needs to know about my health insurance. As for taxation data, including income data – no thank you! (And, of course, while there are plenty of poor people in the U.S., they are not "officially registered" – nor, I imagine, would anyone be eager to have that registration included in their identity card.)

Danish students and faculty draw the line quickly at *religion*. This is in keeping with a strong Danish sensibility – encoded in Danish privacy laws (and, for that matter, in those of the European Union) – that insists on a (more or less) absolute freedom of belief and viewpoint in matters of political ideology and religion. But if we are to enjoy such freedom (as we will explore more fully below), our beliefs and viewpoints must be protected as *private* information.

What about the Thai people? Roughly speaking, while there is a strong opposition among some activists and academics to the national "smart ID card," they have been accepted by the majority of the population as necessary – in part, for example, as such cards, the government has argued, will help in the fight against domestic terrorism.

Overall, then, there emerge these points along a continuum of possible responses:

NO info . . . Minimal info . . . Moderate info . . . Maximum info
(Denmark)   (U.S.)   (Thailand)

Given this continuum and set of points for the sake of specific national/cultural references, where have you and your cohorts drawn the line?

So far as you can tell at this point, how might your sensibilities regarding privacy be connected with the larger national, political, and cultural environments in which you find yourselves?

Interlude: how can we meaningfully talk about 'culture'?

Q: How do you tell the difference between an introverted Norwegian and an extroverted Norwegian?

A: The extroverted Norwegian looks at *your* shoes when he's talking to you . . .

This joke was told to me by Johnny Søraker in 2005, in response to a joke I passed on from Minnesota: "Did you hear about the Norwegian man who loved his wife so much he almost told her?" Both jokes trade on the cultural stereotype of Norwegians as very reserved; both are funny, in my view – especially if they are told by Norwegians (or their descendants) as a way of poking fun at their own tendencies and habits.

These jokes help make a larger point: we recognize that there are sets of behavior patterns (beginning with language), norms and values, preferences, communication styles, and so forth, that are characteristic of one group of people in contrast with others. For at least a century or so, anthropologists and other social scientists have accustomed us to thinking of these sets in terms of "culture." We have just seen an effort to associate specific attitudes and beliefs regarding *privacy* with larger (primarily *national*) cultures. We will continue to pay attention to culture in these ways throughout this volume, but it is important to make clear from the outset how far such references to "culture" are useful, and in what ways these uses of "culture" are limited and, indeed, potentially misleading, if not simply destructive.

To begin with, as I hope these Norwegian jokes suggest, such generalizations about (national) cultures contain at least some grains of truth. In this case, that is, it seems safe to say – as a *generalization* – that indeed many (if not most) of the people born and raised in Norway are, in comparison with, say, the average Midwestern American, more reserved and introverted. Such generalizations are useful – not only for the sake of good humor, but also as starting points for thinking through our differences and similarities. Indeed, for many (most?) people, our culture (however difficult it is to define) usually serves as a core component of our identity, one that demarcates in various ways how we are both alike (in relation to those who share at least many of the elements of the same culture) and different (from those shaped by different cultures).

So, for example, as a Midwesterner, I know that (most of) my East Coast friends will speak and walk more quickly than is the norm in Springfield, Missouri. It is also the case, of course, that these sorts of differences are the occasion for our judging (or, as frequently happens, misjudging) one another on the basis of what is normal (in at least a statistical sense) for our own culture. As in many (most?) other small towns in the U.S. Midwest, the norm is to be "friendly" with, e.g., cashiers and sales clerks, so as to spend a little time in conversation during the course of an otherwise commercial exchange. This friendliness is often (mis)interpreted as time-wasting superficiality by some of my East Coast friends. In turn, their tendency to avoid such small talk might tempt a Midwesterner to (mis)judge them as abrupt, unfriendly, aloof, perhaps arrogant.

And, of course, as we move across national cultures, the differences become even more striking. So as the jokes above suggest, Norwegians tends to be much more reserved, for example, than Southern Europeans. And so on.

I hope these examples illustrate three critical points that should be kept in mind any time the word "culture" appears in this text. First, up to a point at least, these sorts of generalizations are useful – perhaps even essential – if we are to understand and communicate respectfully with one another. That is, if I understand these sorts of cultural differences, I can better anticipate how to communicate appropriately with those who do not share my own cultural values and communicative preferences. So, for example, I am less likely to misinterpret my East Coast friend's curt response (curt, that is, as compared with the norm for a Midwesterner) as rude or unfriendly, and more likely to understand it as intended, i.e., as efficient, to the point, and thereby respectful of our time as a limited and thus valuable commodity. More broadly, these differences are interesting and enriching, as they make us aware of what deeply shapes our individual identities and group norms, and thereby of the incredible richness and diversity of the human family. In particular, these generalizations should thus be helpful to us in coming to understand both ourselves and the multiple Others around us, as we are both similar and irreducibly different in critical ways. Doing so, finally, is necessary if we are to overcome the twin dangers of ethnocentrism (assuming our own ways of doing things are universal) and then judging Others as inferior because their ways are different from our own. Human history is too full of the sorts of warfare, colonization, enslavement, and imperialism that follow upon such ethnocentrism. As Ames and Rosemont put it: ". . . the only thing more dangerous than making cultural generalizations is the reductionism that results from not doing so" (1998: 20). That is, as risky, difficult, and inevitably incomplete as our attempts to characterize culture may be, it seems a necessary exercise if we are to avoid assuming that all others must be like us, and that they are less than fully human if they are not.

But, second, when we use such generalizations, we obviously run the risk of turning them into simple and unfair stereotypes that may foster unjust prejudices rather than intercultural understanding. Hence it must be remembered: every generalization, most especially the generalizations that we think may help characterize a given "culture," by definition entails multiple exceptions to the general rule. In statistical terms, there are always "outliers" – those persons who stand outside the statistical norm as defined by the standard bell curve. So, while many Midwesterners may seem friendly, open, and extroverted as compared with many Norwegians, *of course* there will be at least a few introverted Midwesterners and at least a few extroverted Norwegians who simply confound the generalization. In other words, we must never mistake a generalization for anything other than a generalization – *not*, for example, some sort of universal category that somehow captures an eternal and immutable essence of Midwestern-ness, Norwegian-ness, etc. This is the difference, in short, between a possibly useful generalization and a potentially misleading and destructive stereotype.

Third, however far we can fairly make such generalizations about a given culture, we must further keep in mind that for every individual who may share such national characteristics, she or he is further shaped by a very complex range of additional differences and variations both within and beyond national categories. Folk in Eastern Oklahoma are clearly distinct from folk in Western Oklahoma, just as people in Aarhus (Denmark) have distinct (and not always positive) impressions of how Copenhageners, while clearly Danes, at the same time differ from them (and vice versa, of course). Immigrant communities are distinct in multiple ways, while simultaneously including people seeking to either assimilate to or hybridize with the larger national culture. Indeed, in any given city, a specific neighborhood features a specific set of cultures or subcultures as affiliated with age, ethnicity, and class. And then, of course, gender – generally – makes a difference as well. Oh yes: all of these change over time, of course – some elements more quickly than others – complicating the picture still further.

All of this means, then, that any generalizations we make about a culture can only be taken as *starting* points – and as *dynamic* concepts (i.e., concepts open to change), not *static* concepts. That is, while potentially useful for our initial reflections and encounters with one another, further exploration almost always leads us to more complex and nuanced understandings. As a result, we will almost always modify and perhaps reject altogether elements of these starting points. In fact, we are about to see an example of this sort of modification shortly, in Soraj Hongladarom's account of Buddhist understandings of the person and privacy – an account that will nicely complicate the basic differences between Thai and U.S. culture that we have started with here (see pp. 53–4).

To be sure, the concept of culture is enormously problematic. Briefly, there is no consensus on a definition of "culture." In fact, a number of experienced and respected scholars and researchers who focus on the intersection of culture, information and communication technologies (ICTs), and communication argue that we should abandon the concept altogether (e.g., Kampf 2008; Macfadyen 2008; Nocera 2008). At the same time, however, some notion of culture – for example, as we will see in chapter 4 as represented by specific elements such as the contrast between high-context/low-content (HC) and low-context/high-content (LC) cultures – remains useful when handled with care.

In our context, then, I believe (*judge*) that the grains of truth underlying cultural generalizations, and their utility in helping us overcome ethnocentrism, come to more respectful understandings of Others as *Other* (i.e., as genuinely and irreducibly different from ourselves), and communicate with one another more effectively, help offset the risks and limitations of such generalizations. But this means that if these generalizations are to indeed be useful, they must be used with great care. To reiterate, our use of them must always be accompanied by the several caveats highlighted here, beginning with the understanding that these generalizations are just that – generalizations immediately limited by important counterexamples. Second, we must further keep in mind that any individual or group is characterized not simply by a given national culture, but also by an additional complex array of multiple and often changing cultures and subcultures; then we ought to be able to avoid the worst dangers of such generalizations – including their becoming misleading, if not simply unfair and unjust stereotypes that only reinforce rather than overcome prejudice and naïve ethnocentrism. Third, we must always keep in mind that these generalizations are only starting points and dynamic concepts – ones that we will inevitably alter and modify in light of additional insight and information.

By keeping these comments and caveats in mind, I hope that readers will never be tempted to mistake what I intend as an initial, dynamic, always incomplete, and exception-laden generalization for a stereotype.

Privacy in the global metropolis: initial considerations

In the developed world, we increasingly *are* the digital information that facilitates our lives and engagements with one another. Luciano Floridi has made this point most strongly: a person

> is her or his information. "My" in "my information" is not the same as "my" in "my car" but rather the same as "my" as in "my body" or "my feelings"; it expresses a sense of constitutive *belonging*, not of *external* ownership, a sense in which my body, my feelings, and my information are part of me but are not my (legal) possessions. (2005: 195)

On first glance, this claim may seem too strong. To be sure, as more and more of our entertainment and communication take place via digital media, and as more and more of our lives are captured in digital form, our "digital footprint" expands dramatically. As a simple but telling measure: when the first hard drives became available for PCs in the early 1980s, a 10MB or, even more staggering, 20MB hard drive seemed to offer more than enough data storage for a lifetime's worth of text files. Two decades later, as the personal computer is increasingly the "digital hub" of a life involving TV, radio, personal photographs, and videos, commercially produced music and movies, emails and web-browsing – and, oh yes, text files – required personal data storage is measured in the hundreds of gigabytes; indeed, at the time of this writing, hard drives with a terabyte (1,000 gigabytes) of storage have become commonplace appliances – soon to be included on notebook computers as well.

But is all of this information primarily our property in the sense of an external, legal possession – and/or is Floridi correct to suggest that at least some elements of "our" information *are* who we are, in the same way as we think of ourselves in terms of our own bodies and feelings, for example?

Floridi's claim becomes more persuasive when we consider how much of our lives in the developed world – beginning with, but by no means limited to, important governmental identity information (e.g., Social Security numbers in the U.S., CPR numbers in Denmark, *Fødselnummer* in Norway, etc.), bank and credit card accounts (e.g., the RIB number in France, IBAN and SWIFT numbers, etc.) and so forth – are digitized, processed, and transmitted electronically. Couple this with the metaphor introduced by James Moor (as we saw in chapter 1) – such information is "greased": it is (almost) as easily copied and transmitted to those we may *not* want to see it as to those we *do* want to see it. As the aptly named phenomenon of *identity theft* suggests, losing these sorts of information about us, what we think of as *private* information, to others may well feel and result in harms more like a direct assault on our own bodies and feelings, rather than solely the theft of external property.

You don't have to be paranoid – but it helps . . .

Whatever our individual ethical assessments and responses to these situations may be, many familar *threats* to privacy are well known. Most of us know, for example, to be careful with passwords to important accounts, with PIN numbers for debit and credit cards, and so forth. Most of us are aware that both commercial and governmental databases containing our personal information are the targets of sometimes successful attacks by hackers: once a database is broken into, others are then able to use this information about us – enacting what is rightly called identity theft not only to take money from our bank accounts and charge purchases to our credit cards, but also thereby, in some cases, to jeopardize our own claims to our own identity.

In addition to these sorts of major but comparatively occasional threats, we are constantly vulnerable when we may think that we are safest – i.e., sitting in front of our *personal* computer, sending information via email, engaging in web-browsing (perhaps for shopping), perhaps doing banking transactions. Especially – but by no means exclusively – users of Windows machines face a constant barrage of Trojan horses, worms, and viruses that can, for example, capture and then transmit critical banking information to a third party. Recently, a Trojan horse called Mebroot was discovered to have infected at least 5,000 machines. The program is picked up as a user browses a particular website that exploits security weaknesses in the Internet Explorer browser. The program installs itself on the Master Boot Record (MBR) of the user's computer, where it more easily escapes detection by anti-virus programs. From here, it then downloads an additional program, called a keylogger, which – unbeknownst to the computer user – records keystrokes, e.g., while entering bank account numbers and passwords – and then sends these back to those controlling the programs (BBC News 2008a). Such a threat is simply one among tens of thousands that can cripple a computer and/or compromise its security.

If you're not paranoid yet . . . terrorism and state surveillance

Many of us are further aware that beyond criminals and hackers, as citizens we face additional threats to our privacy – e.g., from corporations who collect, for example, data on individual purchasing choices (usually by consent in exchange for modest discounts or other economic incentives). Governments may be (somewhat ironically) the worst culprits. On the one hand, the modern liberal state exists to protect basic rights – including rights to privacy; but to protect our rights – especially so-called *positive* or entitlement rights, e.g., to education, health care, disability assistance, family benefits such as child support and salary offsets for maternity and paternity leave, etc. – governments clearly require a great deal of personal information about us. How governments *ought to* and actually *do* protect that information from illicit and potentially devastating use against their own citizens varies widely from country to country. Somewhat more darkly, especially following the September 11, 2001, attacks in the United States, governments throughout the world justify ever-greater surveillance of their own (and other) citizens in the name of fighting terrorism. And so unknown (because secret) quantities of personal information – as transmitted through emails, phone calls, faxes, etc. – are collected and scrutinized for potential threats. By the same token, surveillance of citizens through security cameras – distributed ever more densely throughout the world – continues to expand.

Indeed, these threats to privacy may come not only by intention, but also through simple error. For example, in October 2007, two CDs containing data relating to Child Benefits (e.g., names of parents and children, dates of birth, addresses, and National Insurance and bank account numbers) of ca. 25 *million* individuals – nearly half of the population of the United Kingdom – were lost in the mail (BBC News 2008b). At the time of this writing, the disks have not been recovered – and so these individuals and their families remain at risk of fraud.

About that phone you're carrying . . .

Finally (for now, at least), digital media beyond computers and computer networks threaten our privacy in other ways. To use an admittedly sensationalistic example: cellphones equipped with cameras for taking both photos and videos can record us without our knowledge in what we would otherwise believe to be private, perhaps even intimate, situations. The resulting videos can then be shared – either simply with friends, or on a far more public site such as YouTube. In a recent Danish case, for example, adolescent boys took cellphone videos of themselves having sex with young girls – who were apparently not aware that the filming was taking place. The primary purpose of the video – later shown to one's friends – is to prove that, indeed, the young man has had sex with a particular girl (Olsen 2007a). While for the male involved, this may seem to be a relatively harmless way of showing off his adventures to his friends, for the young women involved, there are serious psychological harms when they learn of the existence of what amounts to a public record of what they had believed to be a private and intimate event (Olsen 2007b).[1] (This use of cellphone cameras may be

---

1 Danes will recognize the source here – *Nyhedsavisen* – as one of Denmark's larger but more sensationalistic newspapers. As such, the newspaper is given to a journalistic approach that favors "moral panics" – i.e., highlighting events and possible consequences that will excite readers' fear (and perhaps a little lust), and thereby stoke interest in the story.
       We will discuss moral panics in chapter 5 more fully, especially as they threaten to set up logical and conceptual frameworks that get in the way of clear ethical analysis. Hence, the example offered here is consciously intended as an example from contemporary experience, intended to spark discussion and reflection – but one that must also be viewed with critical skepticism if we are to move beyond sensationalism to more robust forms of ethical reflection.

compared with the still darker practice of "happy slapping" [Kahlweit 2007; Picturephoning.com n.d.]. Cf. the opening discussion, chapter 5.)

More broadly, Theptawee Chokvasin (2007) has pointed out how mobile phones are uniquely personal devices that thereby create a distinctive new communicative space. Briefly, physically fixed landline phones allow for the possibility of our *not* being "there" to answer the phone – either because we are indeed physically away from the phone, and/or as we may have an assistant who is responsible for taking our calls, determining who gets to speak with us when, etc. By contrast, we carry mobile phones with us more or less everywhere, including those places where we otherwise would not have a landline phone. (How many of us have overheard someone carrying on a conversation while using the toilet in a public restroom?) This inverts our earlier context, in which a ringing but unanswered phone could signal the real unavailability of the person we wanted to speak with: with cellphones, everybody knows that if we do not answer a call to our mobile phone, it may well be because we *choose* not to, even though we're physically available to do so. What this means, Chokvasin points out, is that in the era of the mobile phone, we are *always* "there": our "default setting" is to be connected to the communication network – and disconnected usually just by choice.

More starkly: as it inverts the earlier context – as being "on the grid" is now the norm, such that being off the grid or offline can signal a positive desire *not* to communicate – the mobile phone has turned traditional notions of "public" and "private" upside down. Earlier, privacy in the form of being "off the grid" of a public communications network was commonplace. And, especially for the sorts of philosophical and political reasons we will explore more fully below, the capacity to be *incommunicado* in this way was seen to be essential to *being human*. First of all, such privacy makes possible the sort of space and time needed for the development of an autonomous self, one capable of reflecting on and carefully choosing among the multiple acts and values available to human beings, both in solitude and in community with others. In this way, privacy is an essential condition for *our* creating our *selves*. Such autonomy, moreover, is not only a necessary condition for our being suited to living and acting in a democratic society; most fundamentally, as modern political theory emphasizes, only such autonomous selves can justify the existence of democratic societies.

By contrast, the mobile phone (along with other digital media) has in effect made *publicity* our default setting. We are always "on the grid" in the developed world – often by choice, to be sure, but not always. This inverts or turns upside down earlier understandings of who we are, of our relationships with others – and thereby seems to require a radical re-evaluation of our earlier ethical and political philosophies.

Privacy: a generation gap?

Again, our ethical assessments and responses to these situations may vary widely. In particular, as suggested by the example of the Danish adolescents who used their cellphones to video-record their sexual encounters, it is frequently noted that younger people seem less concerned about protecting their privacy, at least as traditionally conceived. Perhaps as having grown up in a cellphone culture – along with the many other digital means of communication that saturate our lives in the developed world – young people have simply had less *experience* of the sorts of privacy available in the pre-digital era?

Along these lines: while 1990s' "web-cam girls" may have begun as yet another instance of how the pornography industry was at the forefront of discerning and developing new possibilities of ICTs – the scenario of allowing others to view the most intimate details of one's private life (at least for a fee) as spectacle quickly became a major form of mainstream entertainment in the late 1990s. While "Big Brother" remains for people in my generation an ultimate icon and trope for a totalitarian regime that works by eliminating all forms of individual privacy, "Big Brother" reality shows that showcase private lives for millions of viewers are rather understood as a form of performance and entertainment. Indeed, for younger people, such unveiling of the personal may be inspired by hope for fame and popularity, if not a quest for authenticity.

But even so, while the ethical sensibilities surrounding privacy may vary in important ways between the generations, younger people clearly do have ethical expectations regarding privacy. Recall the example in chapter 1 of Facebook introducing the Beacon program in the fall of 2007: many of Facebook's largely younger clientele expressed outrage when confronted with what they considered to be a violation of their privacy. As well, it may be fair to say that our under-standings of what kind of privacy we can expect in the online world are gradually catching up with the realities (of, more or less, none). Again, on social networking sites such as Facebook and MySpace, young people often include sexually suggestive photos and com-ments as part of their homepage or profile. Such displays are largely intended to capture attention and increase one's popularity – again, a form of self-revelation that seems to head in the direction of "Big Brother" sorts of celebrity and entertainment. But as young people are also quickly learning – sometimes, the hard way – what they believe to be (at least relatively) private information is oftentimes far more public than they would like: more than one young graduate has discovered that s/he has not been hired into – or, perhaps, that s/he has been fired from – an attractive and enjoyable job because of pho-tographs or other material on their profile that the (prospective) employer found to be inappropriate in one way or another.

In the U.S. context, recent research by the Pew Internet and American Life Project suggests – surprise! – a complex picture regarding teenagers and privacy expectations. On the one hand, the majority of teenage users take steps to protect their privacy – e.g., by limiting access to their profiles, using fake names or first names only, etc. Nonetheless, a majority further believe that a sufficiently interested person would be able to find them through the informa-tion they provide online (Lenhart and Madden 2007: i–iv).

In sum, while the generations may disagree on the nature and limits of privacy, it is clear that we all nonetheless expect, and, in some cases at least, require, some form of privacy and data privacy protection.

For us, then, the primary *ethical* question is: especially as you (increasingly) *are* your data, what sort of "privacy," if any, *ought* you to have?

Privacy: from human universal to cultural differences

On the one hand, it is arguable that "privacy" may count as a value/ notion that is found in all cultures. (See Soraj Hongladarom [2007a: 110f.] for discussion of the work of James Moor [1997, 2002] and Adam D. Moore [2008] on this point.) Crucially, what we understand "privacy" to mean, however, varies widely. Very briefly, in the modern West, "privacy" is considered to be one of the basic *rights* of *individuals* that is to be protected by modern liberal states. Even within the modern West, however, the ethical justifications we offer in support of privacy – and, with it, data privacy protection – vary from country to country. So, for example, in the United States, as Deborah Johnson (2001) has pointed out, we justify privacy as both an *intrinsic* good – i.e., something that we take to be valuable in and of itself – and as an *extrinsic* or instrumental good – i.e., something valuable as a *means* for another (*intrinsic* or *extrinsic*) good.[2] Along these lines, we *need* privacy if we are to become *autonomous selves*: that is, we need privacy to cultivate and practice our abilities to reflect and discern *our own* ethical and political beliefs, for example, and how we might enact these in our daily lives. Privacy is thus a means for the autonomous

whole, good

---

2 We easily recognize that some things are valuable primarily as they serve as *means* to other goods or ends: so, commonly, many students value their education as an *extrinsic* good – i.e., something that is valuable as a means to achieving some other good, such as a job, a good salary, etc. But these in turn may be simply *extrinsic* goods – i.e., goods that are likewise valuable not so much in themselves (e.g., few of us – unfortunately – think of our work as an *intrinsic* good, as something worthwhile in itself, whether or not we are paid for it). So it seems that, somewhere, the chain of justifications for extrinsic goods must come to a rest at an *intrinsic* good – something that is simply worthwhile in and of itself. Or else, as Aristotle famously argued, we are faced with an infinite regress of an extrinsic good being justified by a further extrinsic good, etc. Then the difficulty becomes one of finding such an *intrinsic* good – indeed, one that all of us would agree is valuable in and of itself. But, as Aristotle further argued, *eudaimonia* – often translated as "happiness," but better translated as "contentment" – is a good we all recognize as *intrinsically* valuable. That is, we may well ask someone *why* they want to attend university – that is, what further good justifies such attendance if s/he believes that attending university is only an *extrinsic* good. But we don't seem to need to ask *why* someone would want to be happy or content: that is, happiness or contentment appears to be good in itself, and thus does not require further justification as a means to some further end.

self to develop its own sense of distinctive identity and autonomy, along with other important goods such as intimate relationships. Only though privacy, then, can the autonomous self develop that has the capacity to engage in debate and the other practices of a democratic society (Johnson 2001: ch. 3). In Germany, rights to privacy are likewise considered as a basic right of an autonomous person *qua* citizen in a democratic society. Privacy is also seen as an instrumental good – primarily, as it serves to protect autonomy, the freedom to express one's opinion, the "right of personality" (*Persönlichkeitsrecht*), and the freedom to express one's will. Finally, privacy protection – specifically, data privacy protection – is seen as a means necessary for the development of e-commerce (Bizer 2003).

By contrast, "privacy" in many Asian cultures and countries has traditionally been understood first of all as a *collective* rather than *individual* privacy – e.g., the privacy of the *family*, vis-à-vis the larger society (Ramasoota 2001: 97f., 100f.; cf. Kitiyadisai 2005). Insofar as something resembling *individual* privacy was considered, such privacy was looked upon in primarily negative ways. For example, Japan's Pure Land (*Jodo-shinsyu*) Buddhist tradition emphasizes the notion of *Musi*, "no-self," as crucial to the Buddhist project of achieving Enlightenment – precisely in the form of the dissolution of the "self," understood in Buddhism to be not simply an illusion, but a most pernicious one. That is, as the elemental "Four-Fold Truths" of Buddhism make clear, our discontent or unhappiness as human beings can be traced to *desire* that can never be fulfilled (because either we will never obtain those objects, or if we do, we will lose them again, especially as time and death take them from us). But such *desire*, in turn, is generated by the self or ego. Hence, to eliminate the unhappiness of unfulfilled/ unfulfillable desire, all we need do is eliminate ego or self. The Buddhist goal of *nirvana*, or the "blown-out self," thus justifies the practice of what from a modern Western perspective amounts to intentionally violating one's "privacy": in order to purify and thus eliminate one's "private mind" – thereby achieving *Musi*, "no-self" – one should voluntarily share one's most intimate and shameful secrets (Nakada and Tamura 2005). Similarly negative attitudes towards individual privacy have marked China for most of its history – in part because of the Confucian emphasis on the good of the larger community (see discussion of Confucian ethics, ch. 6). Hence, until only relatively recently, the Chinese term correlating with individual "privacy" (*Yinsi*) held only negative connotations, i.e., of a "shameful secret" or "hidden, bad things" (Lü 2005). Finally, a similar emphasis on community is apparent in many indigenous traditions. So *ubuntu*, as we saw Dan Burk characterize it at the beginning of this chapter, understands personal identity as "dependent upon and defined by the community" – in part, as we will see in more detail in chapter 6, as this African tradition shares with Confucian thought an understanding of the individual as a *relational* being, i.e., as defined by the multiple relationships with others in the larger community. In this light, it makes sense that

> Within the group or community, personal information is common to the group, and attempts to withhold or sequester personal information are viewed as abnormal or deviant. While the boundary between groups may be less permeable to information transfer, *ubuntu* lacks any emphasis on individual privacy. (Burk 2007: 103)

At the same time, however, these understandings of privacy are undergoing significant change. This is in part because globalization, as itself driven by the rapid diffusion of digital media, often thereby increases our awareness of and interactions with one another cross-culturally. This in turn leads to a *hybridization* of diverse cultural values and practices. In particular, as young people in Asia enjoy a growing material wealth and thereby a growing *physical* personal space (i.e., their own room in a family dwelling – something more or less nonexistent a few decades ago), and as they are ever more aware of Western notions and practices regarding individual privacy, thanks to global media, they increasingly insist on *personal* and *individual* privacy in ways that are baffling (at best) and frustrating (at worst) to their parents and their parents' generation (e.g., Lin and Henkes 2004; Kitiyadisai 2005; Lü 2005).

Finally, Soraj Hongladarom (2007a) points out that while earlier cross-cultural discussions of privacy tended to emphasize these sorts of contrasts, there are also important similarities between, say, Western and Buddhist views. To begin with, Buddhism must emphasize at least a relative role and place for the individual: while from an ultimate or enlightened standpoint, the individual is a pernicious illusion, prior to such enlightenment, the individual remains squarely responsible for his/her Enlightenment. For its part, Western thought – both in pre-modern traditions such as Aristotle and in modern philosophical streams such as Hegel – includes emphasis on the community, not simply the individual. From this perspective, Hongladarom has argued for a Thai conception of *individual privacy* – one that ultimately disagrees with Western assumptions regarding the individual as an absolute reality, but nonetheless retains a sufficiently strong role and place for the individual. Such a Buddhist individual, again, is the agent of its own enlightenment, but also serves as a citizen of a fledgling democratic state in Thailand. In this way, Hongladarom argues, there are strong philosophical grounds for granting such an individual privacy rights similar to those enjoyed by Westerners – even if, by comparison, these rights will be more limited in light of the greater role of the state and greater importance (on both Buddhist and Confucian grounds) of the community.

(In other terms that will become important, Hongladarom hereby articulates for us an important *ethical pluralism* regarding the nature of *privacy*. Such a pluralism, as we will explore more fully in chapter 6, stands as a middle ground between ethical *relativism* and ethical *absolutism*. For our purposes here, the important point is that in such pluralism, it is possible to hold together both shared norms and values – in this case, privacy – while these norms and values are *understood, interpreted,* and/or *applied* in diverse ways, i.e., in ways that reflect the distinctive values and norms of diverse cultures. In this way, pluralism allows for a shared global ethics, on the one hand, while avoiding, on the other hand, a kind of *homogenizing* ethics that ignores or obliterates all important cultural differences. And so, ethical pluralism provides the possibility of a global ethics made up of shared norms and values, while preserving the essential differences that define diverse cultural identities.)

Further complications: U.S. vs. E.U. approaches to data privacy protection

Even within (traditionally understood) Western countries, there are important differences in our assumptions and approaches to privacy and data privacy protection. Very briefly, the European Union has encoded in law since 1995 very strong personal data privacy protections (European Union 1995). The E.U. Data Privacy Directives define what counts as personal and sensitive information (e.g., not simply name and address, but also regarding health status, religious and philosophical beliefs, trade union membership, and sexual identity), and require that individuals be notified when such information is collected about them. Individuals further have the right to review and, if necessary, correct information collected about them. As Dan Burk (2007: 98) emphasizes, individuals have the *right to consent* – they must agree to the collection and processing of their personal information. Finally, the Directives insist that when such personal information is to be transferred to third parties *outside* the E.U., such transfers can occur *only if* the recipient countries provide the same level of privacy protection as encoded in the E.U. Directives.

(As Dan Burk characterizes it, in terms that may now be familiar to you from chapter 6, the E.U. approach is strongly *deontological*: it rests upon a conviction that privacy is an *inalienable* right – one that states must protect, even if at considerable economic and other sorts of costs.)

In the United States, by contrast, data privacy protection is something of a patchwork. In general, national or federal regulations address privacy issues with regard to health matters (e.g., the HIPAA [Health Insurance Portability and Accountability Act, 1996, <http://www.hhs.gov/ocr/hipaa>]) and some financial information (e.g. banking and credit information), leaving the rest to either individual states and/or businesses to work out (the latter through so-called "aspirational" models of good practice – see Burk 2007: 97). The default setting here is exactly opposite that of the E.U. model: rather than asking individuals to "opt-in" to having their information collected, processed, and distributed in specific ways, the U.S. approach requires individuals to "opt-out" if they have reservations about how information about them is being collected and possibly used (Burk 2007: 97).

(As Burk goes on to observe, this "business-friendly" approach appears to be the result of a *utilitarian* approach to the issues of data privacy protection. Simply put, the U.S. preference is for minimal governmental involvement and maximum freedom for businesses, in hopes of minimizing the economic – and other – costs of implementing and enforcing more rigorous data privacy protections, such as those of the European Union, and thereby maximizing business efficiencies and profitability. Presumably, doing so will lead to the utilitarian goal of realizing the greatest good for the greatest number – at least in terms of economic gains and benefits. See Burk 2007: 98f.)

As we will see in the next chapter, on matters of copyright, these sharp differences between U.S. and E.U. approaches result in a kind of international competition with regard to whose conceptions of privacy and data privacy shall prevail. In Burk's view, currently, the E.U. approach is spreading "virally" – precisely because the E.U. Directives require those doing business with the E.U. to observe similar data privacy protections (2007: 100f.)

Philosophical considerations

Not surprisingly, there is considerable debate among philosophers in Information and Computing Ethics regarding the nature of privacy, its possible justifications, justifications for its protection, etc. (see, e.g., Floridi 2006; Tavani 2007: 132–4; Tavani 2008). This is not simply an artifact of the dictum that wherever we find two philosophers, we will find at least three opinions; it is further a reflection of the reality that digital media develop and diffuse much more rapidly than our ethical reflections and debates. The following suggestions, then, are contested in the philosophical literature – but to repeat Socrates' admonition: whether we find ourselves in a swimming pool or an ocean, we must start swimming nonetheless.

Herman Tavani (2007) helpfully summarizes three basic kinds of privacy. The first of these is *accessibility privacy* (freedom from unwarranted intrusion). This notion of privacy, also formulated as the right to "being let alone" or "being free from intrusion," is defended in a landmark paper by Samuel Warren and Louis Brandeis in 1890 – who thereby make the first explicit claim in the United States that privacy exists as a legal right (Tavani 2007: 130). Second, *decisional privacy* is defined as a freedom from the interference from others in "one's personal choices, plans, and decisions" (Tavani 2007: 131). Such privacy, Tavani points out, has been crucial in the U.S. context in defending freedom of choice regarding contraception, abortion, and euthanasia (2007: 131). Finally, *informational privacy* is a matter of our having the ability to *control* information about us that we consider to be personal (2007: 131).

To be sure, philosophers such as James Moor (2002), Luciano Floridi (2005), and many others continue to develop important theories that help refine our understanding of what privacy may be and what privacy rights may be justified, especially in the digital age (e.g., Floridi 2008; Tavani 2008). In the meantime, however, we can start our own reflections using these initial distinctions.

What is "private" – and why should it be protected? Up to a point (perhaps), we can distinguish between what we take to be private *situations* and correlative *expectations* regarding privacy, on the one hand, and the (usually digital) *information* that may be collected about those situations, on the other. So, for example, I may find myself at a party with friends, and, after a drink or two, feel inspired to sing some favorite songs. However bad my singing may be, I presume that the party is a private one, and that my friends will have the grace not to share information about my musical incompetence with a larger audience. Similarly, to use the situation suggested in the Danish example: ordinarily (though with plenty of important exceptions), our sexual engagements take place in private places. (Indeed, Moore suggests that this is one of the cultural universals of privacy – i.e., a context or situation in which, in all cultures, privacy is expected.) Again, we (ordinarily) don't expect that *information* about those engagements will be shared with a larger audience.

And there are plenty of other places and contexts where we *expect* privacy – e.g., the therapist's office, in the confessional, and so forth.

By contrast, there are public spaces where we may *not* expect privacy – walking in the park, watching an athletic event in a stadium, and so forth. Unfortunately, there are also gray areas that seem to occupy an in-between space between the clearly public and the clearly private. Sitting at an outside table at a café seems fairly public, and, at least in the U.S. context, I shouldn't be surprised if someone overhears my conversation or perhaps even takes a photograph of the café that accidentally includes me. But what if we're a private party, seated in a room reserved for us at a restaurant?

However we decide about what spaces and contexts are private (or, perhaps more importantly, that we can *expect* to be treated as private), *information* about us and our activities in these spaces adds a new layer of difficulties. I may not mind singing badly for my friends, but I would mind very much if someone recorded my performance and replayed it later for others; or, worse, if I were to discover a video of my singing on YouTube.

This distinction suggests that, at least in those contexts and spaces where I can legitimately expect privacy, I should also be able to control the *information* about my behaviors in those spaces. That is, if I have a right to *accessibility privacy* – a sense that others cannot legitimately intrude upon me and perhaps others in certain contexts – then it would seem that I have a right to *informational privacy* as well.

Even more dramatically, if, as we saw at the outset, we *are* our information (i.e., as Floridi has argued), then it would seem that our rights to control *our* information are even more extensive and important. In other words, insofar as Floridi is right about this, then information about us is more "ours" in the way that *our* bodies and our feelings are *ours* – not simply a more external sort of possession like a computer or a bicycle. This would mean, then, that a violation of our informational privacy – e.g., in the form of *identity theft* – harms us more completely and profoundly than does relatively simple property theft. To put it simply: as bad as it is to have one's possessions stolen, they can be replaced (more or less). But to have our informational identities violated and stolen is worse: it is not at all clear that we can go out and buy a replacement identity in the way we can buy a replacement bicycle. Given that the harm is greater – indeed, that it is a harm directly against the *person*, not simply the person's property – then protecting our informational privacy is even more important than protecting our property. (For these and other sorts of reasons, in fact, both Floridi and Tavani have argued for stringent [informational] privacy rights and data privacy protections.)

Initial reflection/discussion/writing questions

1. How would you define "privacy"?

    It may be helpful here to think of what sort of "things" – acts, events, behaviors, internal notions, imaginations, etc. – you think of as "private," and which sorts of "things" you consider to be "public."

    As a starting point, it may be helpful to review this selection from the E.U. Data Privacy Directives, as a detailed listing of what sorts of "things" are considered *private* and thus as protected information:

    > . . . it is forbidden to process personal data revealing racial or ethnic origin, political opinions, religious or philosophical beliefs, trade-union membership, and the processing of data concerning health or sex life.

2. Discuss as clearly and precisely as possible:

    (a) What kind(s) of privacy do you believe to be most important – especially in terms of the three sorts of privacy described by Tavani?

    (b) Given your account of privacy, do you want to *justify* privacy as either an *intrinsic* and/or *extrinsic* good?

     And: if *extrinsic* – then what is privacy "good for" – that is, for what other (and, ultimately, *intrinsic* goods) does it serve as a means?

    (c) What additional sorts of *justification(s)* can you provide for privacy as you have defined it?

3. Can you discern how far your approach to privacy is shaped by *utilitarian* arguments (such as those at work especially in the U.S. context) and/or by *deontological* arguments (as more characteristic of E.U. approaches, for example?)

4. We have seen that Soraj Hongladarom argues for a Thai notion of privacy that rests on especially Buddhist understandings of the self. As we might expect, Hongladarom goes on to further argue for correlative data privacy protections – protections that might seem limited as compared with contemporary Western (especially E.U.) laws, but are nonetheless recognizable as protections justified for the sake of participating in democratic governance, for example.

But Hongladarom goes further. He draws on the Buddhist analysis of human discontent as rooted in the ego-illusion to point out:

> Violating privacy is motivated by what Buddhists call mental defilements (*kleshas*), of which there are three—greed, anger, and delusion. Since violating privacy normally brings about unfair material benefits, it is in the category of greed. In any case, the antidote is to cultivate love and compassion. Problems in the social domain, according to Buddhists, arise because of these mental defilements, and the ultimate antidote to social problems lies within the individuals themselves and their states of mind. (Hongladarom 2007: 120)

In other words, from a Buddhist perspective, if we want to enjoy privacy protections, then we must go beyond (*negative*) laws that largely tell us what *not* to do (most simply, don't violate others' rights to privacy) to important *positive* ethical injunctions that tell us what *to* do – namely, to pursue enlightenment (in the form of overcoming the ego-illusion), in part through cultivating love and compassion for others.

As we will explore more fully in chapter 6, this recommendation is characteristic not simply of Buddhism, but of *virtue ethics* in the Western tradition. (It further resonates, of course, with "the Golden Rule" – in Christian formulation: do unto others as you would have them do unto you. But of course, the Golden Rule is central to the three Abrahamic faiths of Judaism, Christianity, and Islam – and, indeed, some argue, is found throughout the world, beginning with Confucian traditions.)

(a) How persuasive (or not) do you find Hongladarom's arguments and recommendations regarding privacy – including the *positive* injunction to minimize greed and maximize compassion?

Whatever your response(s), be as clear as you can about your *arguments/evidence/reasons* and/or other grounds for your response(s).

(b) As we have seen, the national and cultural traditions surrounding us have a significant influence on our ethical values and approaches to ethical decision-making. How far can you trace your dis/agreements with Hongladarom to the cultural and national traditions that have shaped your ethical views?

That is, if you *agree* with Hongladarom, is this solely because you likewise have grown up in and remain convinced of the truths of Buddhism? And/or: if you *disagree* with Hongladarom, is this solely because you have grown up in and remain convinced of the truths of other traditions?

And/or: can you find other reasons/grounds/evidence, etc. for your dis/agreement(s) with Hongladarom beyond those reasons, etc. that may hold legitimacy primarily in one culture, but not another?

Additional questions for reflection/discussion/ writing

As we will see in the next chapter, *New York Times* technology columnist David Pogue (2007a) believes that there is a serious "generation gap" with regard to the ethics of copying. In this chapter, we have seen the suggestion that there may be a similar "generation gap" with regard to expectations and beliefs about privacy. That is, younger people may actively pursue self-representation and "consumption" of others as re-presented through various media, perhaps as forms of self-expression and a pursuit of authenticity – constituting a "Big Brother" phenomenon that for older generations conjures up terrifying loss of critical freedoms in an authoritarian regime, rather than simple entertainment, etc.

A. Given your location and experience, do you see evidence for such a "generation gap" in the ethics of privacy and data privacy protection? If so, identify and describe this as fully as possible.

B. Insofar as such differences exist – what do they *mean*?

Here you'll need to review the discussion in chapter 6 on the meta-ethical positions of *ethical relativism, ethical absolutism,* and *ethical pluralism.*

After doing so, do you think/feel that these differences mean largely:

- [relativism] there are no universally valid values, practices, etc. regarding privacy – just as there are no such things for any other ethical matter; all such norms and practices are legitimate solely in relation to a given individual or culture;
- [absolutism] there *are* universally valid values and practices – and so either the older people are right and those who disagree with them are wrong, or the younger people are right, and the older people are wrong;
- [pluralism] both views may be legitimate – as interpretations/ applications of a *shared* set of norms and values regarding privacy.

Whatever your response here, justify it as best you can.

What additional privacy issues do you see evoked by digital media – and how do you respond to them?

Suggested resources for further research/reflection/ writing

Tavani, Herman. 2007. Privacy and Cyberspace (ch. 5). In *Ethics and Technology: Ethical Issues in an Age of Information and Communication Technology* (2nd ed.), 127–68. Hoboken, N.J.: John Wiley and Sons. [An extensive and careful treatment of privacy issues, including attention to issues of data mining and Privacy-Enhancing Tools (PETs), coupled with excellent case-studies, discussion questions, and suggestions for further reading.]
van den Hoven, Jeroen. 2008. Information Technology, Privacy and the Protection of Personal Data. In Jeroen van den Hoven and John

Weckert (eds.), *Information Technology and Moral Philosophy*, 301–21. Cambridge: Cambridge University Press.

[A careful philosophical analysis of why data privacy is important, with particular attention to the debates on these matters between political liberals and communitarians. Van den Hoven specifically takes up the modern Western (liberal) conception of self as a moral autonomy that we have seen to be central, and offers new ways of thinking about data privacy protection in the form of "justified and implementable deontic constraints on flows of personal data" (p. 319).]

See also: Privacy in Cyberspace (ch. 4). In Richard A. Spinello and Herman T. Tavani (eds.), *Readings in Cyberethics* (2nd ed.), 397–500. Boston: Jones and Bartlett, 2004.

[Spinello and Tavani collect here essays by James H. Moor, Dag Elgesem, Helen Nissenbaum, and several others who have made significant contributions to the philosophical discussion of privacy and data privacy protection.]

Nagenborg, Michael. 2005. *Das Private unter den Rahmenbedingungen der IuK-Technologie: Ein Beitrag zur Informationsethik [The Private Sphere under the Framing Conditions of Information and Communication Technologies: A Contribution to Information Ethics]*. Wiesbaden: Verlag für Sozialwissenschaften.

[A very useful – and very extensive – investigation into the multiple ways in which contemporary ICTs threaten classic foundational rights such as freedom, autonomy, equality, concluding with an appeal to develop a consensus-based understanding of privacy and thereby data privacy protection.]

CHAPTER 3

Copying and Distributing via Digital Media

Copyright, Copyleft, Global Perspectives

By posting User Content to any part of the Site, you automatically grant, and you represent and warrant that you have the right to grant, to the Company an irrevocable, perpetual, non-exclusive, transferable, fully paid, worldwide license (with the right to sublicense) to use, copy, publicly perform, publicly display, reformat, translate, excerpt (in whole or in part) and distribute such User Content for any purpose, commercial, advertising, or otherwise, on or in connection with the Site or the promotion thereof, to prepare derivative works of, or incorporate into other works, such User Content, and to grant and authorize sublicenses of the foregoing.

(Facebook Terms of Use [November 15, 2007]:
<http://drury.facebook.com/terms.php>)

We want fans to enjoy their iPods, CD burners, and other devices, but we want them to do so responsibly, respectfully, and within the law.

(Recording Industry Association of America [RIAA], FAQ:
<http://www.riaa.com/faq.php>)

Free software is a matter of liberty, not price. To understand the concept, you should think of free as in free speech, not as in free beer.

(The Free Software Definition [Richard Stallman/GNU Operating
System]: <http://www.gnu.org/philosophy/free-sw.html>)

Chapter overview

We begin with "first thoughts" – a set of reflection/discussion/ writing questions intended to help gather initial thoughts and sensibilities regarding the issues and arguments surrounding matters of copying various forms of digital media. This will also introduce us to the important logical matters of *analogy* and *questionable analogy*.

I then describe the U.S. and European approaches to copyright law and their important ethical differences, followed by discussion of the so-called "copyleft" approaches and important examples of their application in the Free/Libre/Open Source Software (FLOSS) movements. A second set of reflection/discussion/writing questions helps to practice applying these diverse approaches in conjunction with the ethical frameworks of utilitarianism and deontology.

Lastly, I highlight and expand upon the cultural backgrounds and diverse cultural traditions at work here – specifically, Confucian thought and the (southern) African framework of *ubuntu*. We conclude with a final set of reflection/discussion/ writing exercises that include attention to the ethical questions brought to the foreground by this attention to culture.

Initial reflection exercises: illegal downloading vs. stealing (cf. Pogue 2007b)

A. A friend tells you about a new band that she really likes to listen to, and says that you'd really like them as well.

The problem is: as a university student, you don't have the money easily available required to purchase the band's new CD from a local music store.

One solution: your friend knows how to disable the anti-theft tags on the CD, so the alarms should not go off as you walk out the door. Since the likelihood of getting caught is thus very low, she suggests that you can simply steal a copy of the CD from the store.

Yes, I know: you know where this is going, but bear with me for a bit.

As you consider the above scenario carefully . . .

1. What seem to be your options?

That is, the scenario suggests an either/or: either you steal the CD and, presuming you don't get caught, get to enjoy some great music for free – or you don't.

These may well be your two primary options, but in undertaking ethical analysis, it is always a good idea to see if we are clear about *all* the realistic options, not just the most obvious ones.

[*Note*: after you've done this exercise, you may want to review section 4 in chapter 6 on feminist approaches to ethics and an ethics of care.]

2. Develop – either individually, perhaps in group discussion, and/or as a class – as fully as you can the *arguments, evidence, reasons,* and/or other grounds that support *each* of the options you describe in (1) above.

3. Given that you have likely described at least two possible options, each with reasonably strong supporting arguments, at this point, can you provide any additional *arguments, evidence, reasons,* and/or other grounds for a specific choice that help justify that choice as the better of the available options?

4. [(*Optional.*) You may want to review at least the first two ethical [frameworks discussed in chapter 6, *consequentialism/utilitarianism* and *deontology*. After doing so, return to the arguments, etc., that you've provided above. Do you notice whether your arguments are more *consequentialist*, perhaps *utilitarian*, and/or more *deontological* in some way?]

O.k., hold those thoughts . . .

B. A good friend of yours is in a band that is struggling to gain recognition and an audience. All the band members are just getting by on their day jobs – the band as such doesn't make enough money to support any of the members full-time.

The band has just produced a new CD, and they're hoping that it will become a major hit. Following a number of well-known bands before them, they offer on their website the possibility of downloading a sample track from their album, but, of course, hope that this will lead to sales of the full album at the going price of US$10.00.

You are no stranger to illegally downloading music from the Internet. But since you want to support the band, you've gone ahead and paid the US$10.00 for your legal copy of the full album.

1. While in this circumstance, you are willing to pay the US$10.00 required for downloading a legal copy of the album, presume that you also think that under some circumstances it's o.k. to download music from the Internet illegally. With regard to the later case(s), what are your *arguments, evidence, reasons,* and/or other grounds for justifying such illegal downloading?

    [*Nota bene*: this question assumes that a strong *ethical* justification is both distinct from – indeed, may override – arguments based exclusively on current law.]

2. Presuming that you've now marshaled some good *arguments*, etc., that justify at least some sorts of illegal downloading, what *arguments, evidence, reasons,* and/or other grounds come into play in the instance of your deciding *not* to illegally download the full album of your friend's band?

3. [(*Optional.*) Again, you may want to review at least the first two ethical frameworks discussed in chapter 6, *consequentialism/ utilitarianism* and *deontology*. After doing so, return to the arguments, etc., that you've provided above. Can you discern whether your arguments are more *consequentialist*, perhaps *utilitarian* and/or more *deontological* in some way?]

C. Another friend who likes the band's sample track offered for free on the Internet asks you if you'd mind making a CD of your copy of the album, so that he can either:

    (i) highlight the band's music at an upcoming party where he's going to provide the music – in part, so that the album might generate a few more sales; and/or

    (ii) make copies of the album to give to friends of his who are also interested in the music; and/or

    (iii) put a copy of the album on his computer so that it is available to others on the Internet, using one of the current peer-to-peer file sharing networks, and/or

    (iv) all of the above.

1. If you think that you might agree to (i), but not to (iii), explain as best you can:

    (a) what the relevant *differences* are between these two scenarios, and
    (b) what *arguments*, etc., you can provide that can justify your ethical position in both cases.

2. What is your response to (ii) – that is, something of a middle ground between (i) using copies of music to help the band, it is hoped, by generating sales, and (iii) making copies freely available to anyone interested on the Internet – which might well lead to a reduction of the band's sales of its new album?

    Again, for our purposes, whatever your response here, what is important is your analysis of the choice/action and the *arguments*, etc., that support it.

3. [(*Optional.*) Again, as with the optional questions above, you may want to review the arguments developed here vis-à-vis the ethical frameworks of *consequentialism/utilitarianism* and *deontology*, if only to discern which set of arguments you tend to use – so far.]

4. It is likely that at least some number of your class would have responded to the scenario described in (A) – i.e., the possibility of stealing a CD from a music store – by arguing that this would not be a good idea. There are at least two likely arguments here: one *consequentialist* (even if the risk of getting caught is small, the *consequences* of getting caught are potentially catastrophic, and so it's better not to take such a chance); and a second, more *deontological* argument (stealing is simply wrong, even if by stealing you might gain something desirable and enjoyable).

    By contrast, there will likely be many members of the class who are perfectly happy to download, say, a song track or two from a famous (and wealthy) artist whose work is distributed by equally well-to-do multinational corporations. Here, at least in my experience, the arguments tend to be primarily *consequentialist* – e.g., the chances of my getting caught are extremely small, and the very modest profit that both the artist and the multinational corporation lose by my not paying for a legal copy will never be missed by either, since both are already so financially well off.

(a) Are there any additional arguments that occur to you and/ or others in your group/class that work to justify *not* stealing in the first case, but *do* justify illegal downloads in the second case?

(b) Given the arguments that you uncover here, do these arguments always derive from the same framework?

Again, it may be that the arguments *against* stealing a physical copy of a CD include *deontological* arguments, while arguments *for* illegal downloading are primarily *consequentialist*.

If this is the case, then the disagreement between these two cases runs beyond the *first-order* level of what we are to do in a particular instance – the disagreement includes a *second-order* or "meta-theoretical" difference as to which ethical framework(s) we are to make use of (i.e., either *consequentialism* and/or *deontology* . . . and/or any of the additional frameworks described in chapter 6).

(c) At this point, it may be sufficient simply to notice these differences, so far as they seem to be at work – and observe that *if* our arguments do derive from different frameworks, then perhaps there is *not* quite the contradiction that may first appear to be the case (i.e., between disapproving of stealing a CD physically, while approving of illegally downloading a virtual copy of one).

That is, if our arguments against and for (respectively) these forms of stealing derive from different frameworks, then to say that there's a contradiction here is like saying that there's a contradiction between the rules of American baseball and the rules of European soccer. This doesn't make immediate sense: it seems rather that because these are two different games played under two different sets of rules, there can be no serious contradiction between them.

While this observation would relieve us of a first-order contradiction, it nonetheless still leaves us with a *second-order* question, namely: how do we justify – or, to use Aristotle's suggestion, *judge* (i.e., use *phronesis*) – using a specific framework in one instance, and another framework in a different instance?

Thoughts?

D. As we proceed in applying familiar ethical frameworks to the ethical challenges evoked by new technologies, we inevitably proceed by way of *analogy*. And so, in the scenarios described above, I have suggested an analogy between physically stealing a copy of a CD from a music store and illegally downloading a copy from the Internet.

Just to make it explicit – an analogy argument based on the above scenarios might look like this:

- We agree that stealing a physical CD from a music store is wrong.
- Downloading an illegal copy of a music album is *like* stealing a physical CD from a music store.
- Therefore, downloading an illegal copy of a music album is also wrong.

But as good logicians know, every analogy runs the risk of becoming a *faulty* or *questionable* analogy. Such an analogy, rather than helpfully lead us to justifiable conclusions, may instead mislead us. Happily, you don't have to be a logician to see how this is so (though it helps enormously). Rather, at least as a starting point for reflection and discussion, we can draw on the idiomatic phrase "comparing apples and oranges." That is, we sometimes recognize rather easily that a given analogy or comparison is in fact false or misleading somehow – in part because the comparison in fact holds together two radically different sorts of things (the apple and the orange).

So, especially if you *disagree* with the conclusion in the above argument – that illegally downloading a copy of an album on the Internet is ethically wrong – you might be able to make your case by *arguing* for one or more important *differences* between the two scenarios that are held together in the analogy argument.

So: *are* there important, ethically relevant differences between these two scenarios – and if so, what are they?

E. David Pogue (2007a) has recently described his experience in attempting to present a range of ethical scenarios regarding copying and downloading to university students – an experience that left him convinced that there is a serious "generation gap" regarding "copyright morality." That is, while, for Pogue, certain copying scenarios were clearly wrong, these same scenarios seemed ethically acceptable to his audience of university students.

Again, Pogue attributed the difference to a generation gap – suggesting, perhaps, that older people have an appropriate and well-developed ethical sensibility regarding the ethics of copying, while younger people (college or university students) lack this sensibility.

By contrast, Heather Johnson (email, July 16, 2008) is *not* convinced that today's students are somehow less ethical than their elders. She points out that both the range of activities regulated by copyright law in the digital era, and the technologies that allow us to copy and use copyrighted materials (i.e., computers, cellphones, MP3 players, etc.), is far broader than even a decade ago. Moreover, young people may come into contact with these technologies far more often and in contexts much more integrated into their daily routines than other generations. Hence we're asking young people to confront an ethical-technological landscape unknown to their elders. On this view, Pogue's comparison is misleading because we simply don't know how the older generations would have responded to these new technological and ethical landscapes.

It's perfectly possible, that is, that both generations are roughly equal in terms of ethical sensibilities, responsibility, etc. – but in the face of the new range of technical and ethical choices encountered by younger people, they may be making choices that are sensible in their context, but which may be questionable to the elders, who don't understand all the ins and outs of the new technologies. (Also keep in mind here the tendency of media to report on new technologies in the direction of moral panics.) In this light, Pogue's comments amount to comparing "apples and oranges" – i.e., a misleading comparison between two generations facing two fundamentally different technological and ethical landscapes.

1. In your view, *is* there a genuine and significant "generation gap" regarding copying of digital materials? That is, do you notice differences of the sort Pogue describes between older and younger people?

2. If so, why do you think these differences exist?

    That is, Pogue seems to attribute these differences to a comparative lack of ethical sensibility and responsibility among younger people. Johnson, by contrast, argues that the ethical and technological landscapes have so changed with the emergence of digital media that whatever differences we may observe between the generations may only rest on these different landscapes.

3. A subsequent reader of Pogue's column argued:

    > The problem with the RIAA [Recording Industry Association of America] is that they don't care about shades of gray. Young consumers find it ludicrous and counterproductive to spend all of their time parsing our actions. That's why it's so much easier for so many people to say, "Forget it: I'm just going to download it for free." (Pogue 2007b)

    Similarly, Johnson points out:

    > I found my students to be extremely receptive to the idea that many forms of illegal downloading are morally wrong, but that it is problematic that the current moral and legal climate has universally condemned *all* forms of downloading. (email, July 16, 2008)

    From your perspective, who more accurately describes your technological and ethical landscapes regarding copying – Pogue and/or the anonymous reader and Johnson?

    Explain and support your point(s) with examples if you can.

4. [(*Optional.*) Insofar as these differences exist – what do they *mean* with regard to ethics?

   (a) For example, some argue that new digital media require a new ethics, while others argue that traditional ethical frameworks and codes can be effectively applied to new media. Do the ethical responses of younger people represent the emergence of a new ethics, one shaped in response to the new possibilities they represent? And/or: is Pogue correct, in effect, to suggest that, especially as judged from a more traditional ethical framework, some younger people are ethically less thoughtful and responsible?

   (b) As discussed in chapter 6, these sorts of differences can evoke three further responses – *ethical relativism, ethical absolutism,* and/or *ethical pluralism.* After reviewing the description and discussion of these three meta-ethical positions, which seems closest to describing how you treat the differences in copyright morality observed by David Pogue?]

The ethics of copying: is it theft, Open Source, or Confucian homage to the master?

1. Intellectual property: three (Western) approaches

As we saw in chapter 1, there are a number of characteristics of digital media that make the copying and distribution of various kinds of information – whether representing software, a text, a song or video, etc. – much easier to do than with comparable analogue media. That is, once we have access to the various components required – an access that in turn is likewise growing rapidly around the world, despite the grave difficulties of the "digital divide" – copying and distributing a file in digital format is both trivially easy and all but cost-free.

Moreover, the general rules, guidelines, and laws applicable to such copying are wide-ranging and frequently shifting. In its on-going battle against illegal music downloading, the entertainment industry relentlessly lobbies for more stringent laws intended to stop (or at least slow down) widespread distribution of music and video files on the Internet via peer-to-peer (p2p) file-sharing networks; these industries have likewise pushed for Digital Rights Management (DRM) and copy protection schemes also designed to prevent illegal copying or piracy – efforts backed in the United States by the Digital Millennium Copyright Act (DMCA). Critics such as the Electronic Freedom Foundation (EFF) counter that the "DMCA has been a disaster for innovation, free speech, fair use, and competition" <http://www.eff.org/issues/drm>.

(a) Copyright in the United States and Europe

The polarities exemplified by the RIAA vs. the EFF in fact entail at least three major positions or streams of response that we can consider as ethical responses to these sorts of dilemmas.

To begin with, as Dan Burk (2007) characterizes it, intellectual property (IP) law in the United States is shaped by a *utilitarian* ethic (see ch. 6 for discussion), one that argues that copyright and other forms of intellectual property protection are justified as these contribute to the larger public good over the long run. That is, proponents of this view believe that authors, artists, software designers, and other creative agents will take the trouble to innovate and develop new products and services that will benefit the larger public only if those agents can themselves be assured of a significant personal reward – primarily in terms of money or other economic goods. What this means in practice, however, is that it is primarily the industries that have a strong economic interest in copyright and other protections that argue and lobby for such protections. Indeed, the interests and possible benefits of the individual agent are secondary in this view. Given its utilitarian framework, "The rights of the author should at least in theory extend no further than necessary to benefit the public and conceivably could be eliminated entirely if a convincing case against public benefit could be shown" (Burk 2007: 96).

By contrast, European approaches to copyright can be characterized as *deontological* in character. As Burk puts it,

> . . . copyright is justified as an intrinsic *right* of the author, a necessary recognition of the author's identity or personhood. . . . the general ratio-

nale for copyright in this tradition regards creative work as an artifact that has been invested with some measure of the author's personality or that reflects the author's individuality. *Out of respect for the autonomy* and *humanity* of the author, that artifact deserves legal recognition. (2007: 96, emphasis added, CE)

As Burk goes on to point out, we are thus currently caught in an international competition between the U.S. and the E.U. as to which of these approaches to copyright will prevail – with the U.S. currently dominating, according to Burk (2007: 99–100).

The third ethical response to these sorts of dilemmas – namely, copyleft/FLOSS – is the subject of the following section.

(b) Copyleft/FLOSS

Third, *alternatives* to what are seen as excessively restrictive conditions especially on the development and use of computer software have been developed – initially under the rubric of Free and Open Source Software (FOSS). More broadly, these comparatively early notions of Free and Open Source Software (FOSS) are now discussed in terms of a more inclusive acronym – Free/Libre/Open Source Software (FLOSS) – in recognition that a great deal of the interest and work in FOSS operates in Latin-speaking countries (primarily Latin America and the Francophone countries).

This rubric, in fact, conjoins two important but conflicting philosophical and ethical frameworks – those of the Free Software (FS) movement affiliated with Richard Stallman and the Free Software Foundation, and those of the subsequent Open Source Initiative (OSI) begun in 1998 by Eric Raymond and others. While both share the common goal of fostering the development of software to be made freely available for others to copy, use, modify, and then redistribute, the Free Software movement began in largely conscious opposition to commercial development of profit-oriented proprietary software and the copyright schemes seen to protect such software. By contrast, the Open Source Initiative was driven by the goal of making free software more attractive to for-profit businesses. These differences, as we will see, are significant for us on a number of levels: we will explore these in greater detail in *Additional topics, questions, resources*, question 1 (p. 97, below). But for now, we can begin to explore FLOSS by way of Stallman's basic definition of what "free" in "free software" means:

> Free software is a matter of the users' freedom to run, copy, distribute, study, change and improve the software. More precisely, it refers to four kinds of freedom, for the users of the software:
>
> - The freedom to run the program, for any purpose (freedom 0).
> - The freedom to study how the program works, and adapt it to your needs (freedom 1). Access to the source code is a precondition for this.
> - The freedom to redistribute copies so you can help your neighbor (freedom 2).
> - The freedom to improve the program, and release your improvements to the public, so that the whole community benefits (freedom 3). Access to the source code is a precondition for this.
>
> A program is free software if users have all of these freedoms. Thus, you should be free to redistribute copies, either with or without modifications, either gratis or charging a fee for distribution, to anyone anywhere. Being free to do these things means (among other things) that you do not have to ask or pay for permission. (The Free Software Definition [Richard Stallman/GNU Operating System], partially quoted at the outset of this chapter: <http://www.gnu.org/philosophy/free-sw.html>)

Ethically, what is interesting here is the justification for such freedom in terms of *benefits to the whole community*. Rather than relying on copyright schemes as oriented either towards economic incentives or protecting authorial rights as the engines of creative development and distribution, the Free Software movement begins with a more communitarian sensibility, one that is inspired in part by a deep conviction that the potential benefits of computer software (and information more generally) should be shared as broadly and equally as possible.

To state this slightly differently, it may be helpful to understand that by "property right" in the contemporary world, we mean first of all a right to *access* and *use* something – whether a material item (such as your pen, backpack, computer, bicycle, etc.) or something more non-material, including "intellectual property" such as an author's words or a computer programmer's code. Given that property means first of such a *right*, then we can distinguish between copyright and copyleft/FLOSS approaches in terms of *exclusive* and *inclusive* property rights. Briefly, the U.S. and European copyright approaches tend to start with and favor *exclusive* property rights. That is, property rights (of access and use) belong *exclusively* to the individual owner: the "default setting" of such exclusive rights is that the owner has the right to exclude others from use and access to his or her property. Copyleft/FLOSS approaches, by contrast, include notions of *inclusive* property rights. So, in Richard Stallman's definition of free software (quoted above), the starting point is the *users'* freedom – i.e., a community of software users – not the individual's right to exclude others from use and access.

Similarly, the Creative Commons approach, while recognizing and protecting individual rights ("some rights reserved"), does so in a way that is *inclusive* – i.e., "by default" recognizes the rights of others to access and use property. So the Creative Commons "Attribution-Noncommercial-Share Alike 3.0 United States" license reads:

You are free:

–   to Share – to copy, distribute, display, and perform the work
–   to Remix – to make derivative works

Under the following conditions:

–   *Attribution*. You must attribute the work in the manner specified by the author or licensor (but not in any way that suggests that they endorse you or your use of the work).
–   *Noncommercial*. You may not use this work for commercial purposes.
–   *Share Alike*. If you alter, transform, or build upon this work, you may distribute the resulting work only under the same or similar license to this one.

    •   For any reuse or distribution, you must make clear to others the license terms of this work. The best way to do this is with a link to this webpage.
    •   Any of the above conditions can be waived if you get permission from the copyright holder.
    •   Apart from the remix rights granted under this license, nothing in this license impairs or restricts the author's moral rights

<http://creativecommons.org/licenses/by-nc-sa/3.0/us/>

That is, individuals retain their "moral rights" – including a right to *exclude* others from using one's property for others' commercial advantage. At the same time, however, *others'* rights of access and use, including copying, distributing, and remixing an individual's property, are likewise granted under this license: that is, the individual owner's rights are in this way *inclusive* rather than *exclusive.*

In these terms, we will see when we focus more on the cultural backgrounds at work here that a wide range of non-Western traditions and approaches to property – beginning with the initial discussion of the (southern) African tradition of *ubuntu* (next section) – likewise stress *inclusive* rather than *exclusive* rights.

FLOSS in practice: the Linux operating system. As is well known among Linux enthusiasts especially, Linus Torvalds developed a variant of the UNIX operating system[1] in the early 1990s that was intended for free distribution from the outset – and this in the Free Software sense: that is, Torvalds chose to distribute his software under Stallman's GNU General Public License. Subsequently, a great deal of FLOSS work focused on the development and distribution of the Linux operating system and affiliated applications. Among the many distributions of Linux that can be freely downloaded online, the Ubuntu distribution is one of the most popular – in part, perhaps, because of its clear communitarian philosophy. Drawing directly upon the philosophical principles of the Free Software Foundation (as articulated by Stallman, above), the Ubuntu distribution website declares: "Our work is driven by a philosophy on software freedom that aims to spread and bring the benefits of software to all parts of the world" (<http://www.ubuntu.com/community/ubuntustory/philosophy>). Indeed, Ubuntu is driven by a specific sort of *philosophy*:

---

1 For those of us who are not computer geeks: the operating system or OS is the base-level software that defines the various operations required to make your computer "work" – including reading and writing files from various media (floppy disks, CDs, DVDs, memory sticks) and through various communication channels and networks (phone lines, Ethernet connection, wireless networks), along with the many operations required to let you interact with and use that information (e.g., keyboards and mice and the computer screen). For most of us in the developed world, this means in practice some variant of Windows or the Macintosh OS X. Application software, by contrast, is software that runs, so to speak, on top of the OS: this commonly includes applications that make possible wordprocessing, email, spreadsheets, presentation, web-browsing, instant messaging, etc.

Ubuntu is an African concept of "humanity towards others". It is "the belief in a universal bond of sharing that connects all humanity". (Code of Conduct: <http://www.ubuntu.com/community/conduct>)

Consequently, Ubuntu Linux explicitly invokes the Free Software Foundation's definition of the freedoms constituting the "free" in free software, including:

> The freedom to redistribute copies so you can help others. [freedom 2]
> The freedom to improve the programme and release your improvements to the public, so that everyone benefits. [freedom 3]
>
> (see "Our Philosophy," <http://www.ubuntu.com/community/ubuntustory/philosophy>)

The clear intersection between Free Software sensibilities and the *ubuntu* tradition, then, is the emphasis on *inclusive* rather than *exclusive* property rights – and for the sake of benefiting one's neighbors and the larger (indeed, now world-wide) community.

Ubuntu as a Free Software project hence directly reflects the greater emphasis on *community well-being* that characterizes indigenous (southern) African cultural values. At the same time, you may remember that in chapter 2 we saw this emphasis on the larger community as a characteristic of other cultural traditions – especially Confucian and Buddhist traditions – and its consequences for non-Western conceptions of *privacy*. We are now starting to see how this emphasis on community well-being has crucial consequences for our notions of property and thereby such common acts as copying and distributing via digital media. We will return to this discussion of culture and ethics in the next section, as we consider Confucian thought and copyright.

FLOSS in practice: applications and Wikipedia. The FLOSS movements have produced not only operating systems such as Linux, but also equally popular web browsers and email clients (perhaps most notably, Firefox and Thunderbird, respectively, which run on Windows and Macintosh as well as on Linux machines) and even complete office suites that largely duplicate the functionalities of Microsoft's Office software (primarily, Open Office). Such software is hence not only attractive to young people and university students with limited resources; it may also play a vital role in efforts to overcome the "digital divide" and to exploit digital media for the sake of development. (See *Resources on FLOSS/ICT4D* at the end of this chapter, p. 103.)

But the ethical sensibilities and developed applications of FLOSS are not limited to computer applications; in addition, they have generated other fruitful, perhaps even essential kinds of sharing online. One example is the Wikipedia encyclopedia project (http://www.wikipedia.org). Exploiting the famous *interactivity* possible on the Web – such that individuals can not simply read, but also actively write for and contribute other forms of media to a given webpage – Wikipedia has become a remarkable resource, representing the contributions of more than 75,000 active contributors, totaling more than nine million articles in 250 languages (see Wikipedia: About: <http://en.wikipedia.org/wiki/Wikipedia:About>). In contrast with traditional copyright schemes, Wikipedia relies on the GNU Free Documentation License (GNU FDL, or simply GFDL), a copyleft license originally developed by the Free Software Foundation (FSF) for documentation affiliated with the GNU project (see <http://en.wikipedia.org/wiki/GNU_Free_Documentation_License>). The license is designed to accomplish two goals:

> The purpose of this License is to make a manual, textbook, or other functional and useful document "free" in the sense of freedom: to assure everyone the effective freedom to copy and redistribute it, with or without modifying it, either commercially or noncommercially. Secondarily, this License preserves for the author and publisher a way to get credit for their work, while not being considered responsible for modifications made by others.
>
> (<http://en.wikipedia.org/wiki/Wikipedia: Text_of_the_GNU_Free_Documentation_License>)

While Wikipedia is clear that it is *not* to be used as a primary resource for academic research (you have been warned!), nonetheless, it has become one of the most common first stops in undertaking research: in particular, because the materials here may be updated and corrected much more quickly than printed sources, Wikipedia articles may be especially useful (at least as a starting point) for looking into current events, recent changes in a field, and so forth.

In these ways, Wikipedia – along with other "products" of the FLOSS movement – may be taken as paradigmatic fulfillments of the philosophical claims and assumptions underlying these alternative or copyleft approaches. And as exemplifying and fulfilling some of the best promises of the FLOSS movement, these examples provide strong justification for the *ethical* frameworks and approaches at work in their licensing schemes. They thereby serve as important counterexamples to proponents of more traditional (either U.S. or E.U.) copyright schemes, especially as such proponents might argue that FLOSS approaches are somehow utopian, excessively idealistic and impracticable, etc.

Reflection/discussion/writing questions: IP, ethics, and social networking

We've now seen a range of possible approaches to how intellectual property may be treated:

(i)   U.S., *property*-oriented copyright law (consequentialist);
(ii) E.U., copyright law, oriented towards *authorial rights* (deontological);
(iii) Open Source/FLOSS/"copyleft" schemes, including Creative Commons and GNU General Public (GPL) and Free Documentation (FDL) licenses.

1. Given your own country/location, which of these licensing schemes seems to be prevalent in your experience?

2. In your view, what are the most important – but especially *ethically relevant* – differences between these three approaches? Be careful here, and, insofar as you are now familiar with one or more of the *ethical frameworks* discussed in chapter 6 (beginning with *utilitarianism* and *deontology*), try to discern how far

    a distinctive ethical characteristic of a given licensing scheme

    may be seen to depend upon

    a given *ethical framework*.

3. Return to your responses to one or more of the scenarios intro-
    duced at the start of this chapter – e.g., stealing a CD from a
    music store, making an illegal copy of new music for a friend,
    making your music library available for others online through a
    p2p network, etc.

    (i) Which of these three approaches to IP seems *closest* to
         your own responses to such scenarios and the *ethical
         justifications* for those responses that you have developed?
    (ii) Which of these three approaches to IP most clearly *contra-
         dicts* your own responses and justifications?
    (iii) Develop a summary of the *arguments, evidence,* and/or other
         reasons offered in support for the approaches you have
         identified in (i) and (ii).
             Now: in light of the contrasts here between the *argu-
         ments, evidence,* etc. can you discern additional arguments,
         evidence, etc. that might support one of these approaches
         more strongly than the other?

4. Presuming you have an account on a social networking service
    such as Facebook, MySpace, and/or others:

    (i) When you signed up for the account, did you review the "Terms
         of Use" or equivalent legal/ethical agreements required of you
         as a user of the site and its affiliated software?
             If so, why? If not, why not?

    (ii) Review the "Terms of Use" for your networking site – looking
         particularly for the important claims it makes regarding your
         ownership of the materials that you post on the site. (For
         Facebook users, the pertinent section of the "Terms of Use" are
         reproduced at the beginning of this chapter.)

    (iii) Are there claims here that

         (a) surprise you,

         and/or

         (b) upon reflection, you may not be comfortable agreeing to?

If so, identify these (both for your own reflection and, perhaps, for class discussion and further writing).

(iv) Can you discern which of the three approaches to IP that we have examined are presumed in these claims?

If so, is part of your discomfort with the claims made upon you here because you have a strong *ethical* disagreement with the approach to IP presumed here? That is, can you *argue* – most easily, from a *different* ethical approach to IP – that the claims made upon you are somehow wrong?

(v) Social networking sites depend on acquiring as many user accounts as possible in order to make money (primarily through advertising, the sale of at least aggregated information about its users, etc.). In this way, they are at least somewhat sensitive to the interests, needs, and opinions of their users.

If you find that the "Terms of Use" of your favorite social networking site conflict with your own ethics and underlying assumptions regarding IP, it would be an interesting exercise to write the site owners (either individually and/or as a larger group) and explain your disagreements and reasons for these.

If nothing else, their response(s) to your communications might provide additional material for interesting ethical analysis!

Additional (meta-ethical) questions

5. As we have seen, digital media make copying and distributing materials – whether computer software or entertainment resources such as songs, videos, etc. – much easier and quicker than with previous sorts of media.

Some thinkers and proponents – including Stallman, Lawrence Lessig, and others we briefly encountered above – have suggested that because digital media are distinctive in these ways, they require either

(i) a very different sort of ethics with regard to IP – but one that still represents an extension of familiar ethical frameworks as applied to digital media, or

(ii) a radically new sort of ethics – perhaps one that we have yet to fully develop and articulate, much less apply carefully to the ethical issues of IP (as well as other ethical issues evoked by digital media).

What is your response to these sorts of claims? That is, do you think that

(a) IP issues can be fully and satisfactorily dealt with using traditional ethical frameworks, and/or

(b) IP (and other ethical issues in digital media) will require totally new ways of thinking ethically?

Try to justify your response here, perhaps by way of using a particular example from either your own experience and/or one or more of the scenarios discussed in this text.

2. Intellectual property and culture: Confucian ethics and African thought

Finally, as the example of Ubuntu and the differences between U.S. and E.U. approaches to copyright suggest, our attitudes and approaches to matters of intellectual property and how far and under what circumstances such materials may be justifiably shared with others are strongly shaped by *culture*. (Keep in mind, of course, the sense and limitations of any generalizations we may try to make about culture: see ch. 2, *Interlude*, pp. 39–43.)

As a further example: U.S. copyright law is moderately clear with regard to what counts as "fair use" for teaching and research purposes – at least as far as printed materials are concerned.[2] In particular, under most circumstances, it is illegal for me to make, say, photocopies of an entire book that I would then distribute to my students at the beginning of the semester for their use during the

---

2 The U.S. Copyright Act of 1976 was accompanied by the development of "Guidelines for Classroom Copying in Not-for-Profit Educational Institutions with Respect to Books and Periodicals" (see, e.g. <http:// www.lib.umn.edu/ copyright/ classguide.phtml>). Perhaps by this point, readers will not be surprised to discover that an equivalent set of guidelines has yet to be established for "fair use" copying of digital materials.

course. On the other hand, in general I am allowed to place original materials, such as articles or book chapters, on reserve for my students in the library; they are then free to check out these materials and make copies of them – as part of their "fair use" of these materials as students.

By contrast, European copyright law makes no equivalent provisions for "fair use." On the other hand, in Thailand I received a now highly cherished gift from some graduate students: a nicely photo-copied version of an important book in philosophy of technology, complete with a carefully crafted cover, on which the students inscribed their names. In U.S. circumstances, this could only be seen as a crass violation of copyright law, but, in the Thai context, this copying was seen to be a mark of respect both for the (famous and well-known) author of the text and for me as the recipient of the gift.

In the latter case, the gift from the students reflected not simply relatively limited economic resources – a (consequentialist) reason often cited as a justification for making illegal copies of materials. In addition, the students' gift reflected the influence of Confucian tradition: as Dan Burk has summarized it, Confucian tradition emphasizes emulation of revered classics – and in this way, copying (as it was for medieval monks in the West) is an activity that expresses highest respect for the work of the author (2007: 101). By the same token, a master philosopher or thinker is motivated primarily by the desire to *benefit* others with his or her work – rather than, say, personally profit through the sale of that work – and so s/he would *want* to see that work copied and distributed widely rather than restricted in its distribution.

In this light, the Confucian tradition and practice thus closely resemble what we have already seen of *ubuntu* as a (southern) African cultural tradition. While they are distinct from one another in crucial ways, they share the sense that individuals are relational beings, ones centrally interdependent with the larger community for their very existence and sense of meaning as human beings. Compared with Western systems that highlight the individual and the individual's *exclusive* property right, both Confucian and *ubuntu* traditions downplay the importance of the individual and individual interests, stressing instead the importance of contributing to and maintaining the harmony and well-being of the larger community. (We will explore these matters more fully in ch. 6, but it is important to stress here that this emphasis on the community does *not* mean – as it sometimes appears to for my Western students – the complete loss of "the individual." On the contrary, individual human beings retain significance and integrity in these views, precisely as they are able to interact with others in ways that foster community harmony and well-being.)

Hence, whether it is copying an important text out of respect and gratitude, and a desire to show respect and gratitude (my Thai students), or making available an operating system such as Ubuntu for free (in more than just the economic sense of being without cost), in both cases, the approach to what counts as property is *inclusive*: the right to access and use these materials belongs to the community, not exclusively to the individual.

In sum: we have now seen culturally variable understandings of property and the ethics of copying and distribution – initially within Western cultures (U.S. and European copyright schemes, along with copyleft schemes affiliated with FLOSS), and now between Western and non-Western cultures and traditions. In this light, it should now be clear that the various software operating systems and applications developed under FLOSS are popular in the developing world not simply for economic reasons: that is, at least in terms of licensing arrangements (though not necessarily in terms of technical and administrative costs), FLOSS avoids the licensing fees characteristically charged by corporations such as Microsoft. In addition, we have seen what we can properly call the *ethos* or ethical sensibilities surrounding FLOSS: this *ethos* includes an explicit emphasis on one's contribution to a *shared* work for the sake of a larger community, and a sense that "information wants to be free" (meaning specifically, freely copied). Moreover, this *ethos* closely resonates with the emphasis on community well-being that we have now seen to be characteristic of Confucian tradition and *ubuntu*, as but two examples of non-Western philosophical and ethical traditions.

And, presuming you read chapter 2 before this one, there is a larger coherency that, I hope, is also beginning to become clear:

just as major cultural variations regarding our understanding of the individual vis-à-vis the community shape our conceptions of privacy and expectations regarding data privacy protection, so these major cultural variations likewise shape our understandings of property and the ethics of copying and sharing.

Reflection/discussion/writing questions: law, ethics, culture

1. Copying: law, culture – ethics?

Does the *legality* of copying music make a difference *ethically?* And: how do our cultural attitudes towards texts, authorship, and property affect our ethical analyses of copying?

We have now seen a continuum of possible approaches to notions of intellectual property and the ethics of copying and distributing such properties. One way to schematize that continuum looks like this:

US/Europe	FLOSS /copyleft	Confucian
		Ubuntu

←————————————————————————→

Greater stress on		Greater stress on
individual		*community*
exclusive property rights		*inclusive property rights*

(Again, these generalizations about culture are starting points only.)

As you review your initial arguments and responses to the questions concerning copying and distributing copyrighted materials:

A. Can you now see one or more ways in which your views, arguments, etc., rested on one or more of the assumptions underlying these three diverse approaches to intellectual property? That is, how far (if at all) do any of your views, arguments, etc., rest on

   assumptions about the relative importance of the individual vis-à-vis the community

and/or

assumptions about the nature of property rights (exclusive or inclusive)?

If so, identify the specific assumption(s) at work in your initial arguments and views.

B. Does it appear that your relying on these assumptions is related to your culture(s) of origin and experience?
That is, do

the assumptions you're making regarding either the individual/community relationship and/or the inclusive/exclusive character of property

correlate/not correlate

with these assumptions as characterizing the larger culture(s) of your origin and experience?

C. Especially if there *is* a correlation between the assumption(s) underlying your views and arguments and the culture(s) of your origin and experience, what does that *mean* in terms of ethics?
This is to say: recognizing the role of culturally variable norms, beliefs, practices, etc., in our ethical arguments characteristically leads to at least two sorts of questions:

(i) Are our ethical norms, beliefs, practices, etc., *ethically relative* – i.e., entirely reducible to the norms, beliefs, practices, etc. of a particular culture? If so, then we could say, for example:

for persons in a Western culture whose basic assumptions tend to support individual and exclusive notions of property and thus more restrictive copyright laws –

if those persons violate more restrictive copyright laws (e.g., through illegally copying and distributing music), they thereby violate the basic ethical norms of their culture and should be condemned as wrong;

but:

for persons in, say, a Confucian culture whose basic assumptions tend to support more community-oriented, inclusive notions of property and thus less restrictive copyright laws –

if those persons violate the more restrictive copyright laws of Western nations, they are thereby simply following the moral norms and practices of their culture, and should not be condemned as wrong.

Consider/discuss/write: Does this approach of ethical relativism to the sorts of differences we have seen "make sense" to you as a way of how we are to understand and respond to these deep differences between cultures?

If so, explain why. If not, why not?

(ii) If you do think there's something mistaken about the above scenario – and, thereby, about *ethical relativism*, then additional questions arise:

(a) Do you want to shift to a posture of *ethical absolutism* – claiming that the norms, beliefs, and practices of country/ culture *X* are the right ones: those countries/cultures/ individuals who hold different norms and beliefs are thereby wrong?

and/or

(b) Do you think it's possible – as we saw in chapter 2 on privacy – to develop an approach to matters of copying and distributing digital media that works as an *ethical pluralism*?

As a reminder: ethical pluralism conjoins shared norms or values with diverse interpretations/applications/under-standings of those norms and values – so as to thereby reflect precisely the often very different basic assumptions and beliefs that define different cultural and ethical traditions.

Consider/discuss/write: Given what we've seen regarding the current conflicts between U.S. and European approaches to copyright law (above, pp. 74–5), do these conflicts point towards an *ethically absolutist* approach on the part of the different countries engaged in these conflicts?

And/or: in light of those conflicts, do you see any possibility of an ethically pluralistic solution emerging?

D. If you find that your beliefs, norms, and practices do *not* correlate with those underlying the culture(s) of your origin and experience, why might this be the case?

Are we – especially in terms of our ethical sensibilities – somehow capable of discerning and establishing moral norms *apart* from, perhaps even *against*, prevailing norms and assumptions of our culture(s) of origin and experience?

If so, how does that "work" in your view? That is, how do we as human beings come to develop our own ethical sensibilities? On what grounds? Etc.

2. Copyright: different ethics for different countries, cultures?

A student from a developing country justified the practice of pirating in that country – of illegal copying and selling imported music CDs – under the conditions that they were

(a) the work of well-to-do (and primarily Western) artists and
(b) distributed and sold in that country by equally well-to-do multinational corporations.

The student justified the practice of pirating in an interesting way:

(i) The widespread practice of pirating – of illegal copying and selling – imported music CDs effected an interesting change.

Originally, imported CDs cost ca. US$10.00. Pirated CDs were being sold for US$1.00. But after a certain period of time, the prices of legal, imported CDs dropped to US$2.00 – thereby making them much more affordable for that country's inhabitants, and thus allowing the multinational corporation and Western artist to make at least more profit than they had before.

This is to say: illegal copying and sales of CDs in effect broke a market monopoly, so that the market forces worked as they are supposed to – i.e., with free(er) competition leading to lower prices.

[*Historical reference*: Adam Smith, the author of the philosophical arguments that ground modern capitalism (*Wealth of Nations*, 1776), justified modern capitalism in clear *utilitarian* terms. Very briefly, free markets that thereby allow for competition should, if coupled with the laws of supply and demand, lead to the economic greatest good for the greatest number: the largest possible supply of goods and services made available at the lowest possible prices for the largest number of people.

While Adam Smith would not countenance breaking the law, this first justification offers a *utilitarian* argument, insofar as it argues that the practice of pirating, in effect, helped the market function as it should, and thereby resulted in at least greater good for more people than if people only had the choice to buy legal imported CDs at US$10.00 each.]

In addition, the student pointed out that, by contrast, many students and others of limited means consciously choose to pay full price for a CD produced by a local/regional/national music group. Again, the argument is, on first blush, *utilitarian*:

(ii) By paying full price for CDs produced by local/regional/ national musicians, they thereby supported those who really needed it – and thereby helped boost their own economy.

In both examples, the student's arguments echo the arguments I hear from many students in the developed world. Again, in the case of nationally or internationally known musician(s), whose work is distributed by wealthy and powerful corporations, the positive benefits or consequences of illegal copying and downloading (in terms of making the music more easily available for more people) outweigh the possible negative costs (of a modest amount of lost profit to the musicians and the companies). By contrast, many will make a conscious effort to "buy local" – to pay full price for CDs produced and distributed by local bands struggling to make a start.

Responses? In particular:

(i) Does it seem to you that, say, students and others in developing countries can make a greater/stronger case for pirating and other forms of illegal copying than students and others in developed countries?

(ii) Assume that the developing country in this example is a country marked by one of the more community-oriented traditions discussed above – e.g., *ubuntu* or Confucian thought.

And assume that the students in the developed world that I refer to live in the well-to-do countries such as the United States and Scandinavia – i.e., countries and traditions shaped by Western conceptions of the individual and primarily exclusive property rights.

In light of the important differences between the cultural and ethical backgrounds, how do you respond to the claim that:

the students in the developing country (shaped by *ubuntu* or Confucian tradition)

have a stronger justification for their illegal copying than

Western students?

Or would you rather argue that everyone should follow the copyright laws – no matter their location and culture?

3. Copyright and deontological ethics

Deontological ethics, as emphasizing, e.g., duties to respect and protect the rights of others – whatever the costs of doing so – can be invoked in these debates as offering reasons for obeying the law. Even if the consequences of doing so may be unpleasant – e.g., not having access to music one would otherwise enjoy – doing so nonetheless reflects an important duty to respect the property rights of others.

Such duties, however, crucially depend on establishing that the laws in question are *just* laws – i.e., grounded in one or more set of values and principles that are used to demonstrate that such laws are justified as means to higher ends.

And so, Mahatma Gandhi and Dr. Martin Luther King, Jr. (and, for that matter, the signers of the U.S. Declaration of Independence) famously argued that while we are morally obliged to follow *just* laws, we are allowed, even morally *obliged*, to *disobey unjust* laws.

The trick, of course, is demonstrating that a given law is indeed *unjust*.

Some arguments I've heard in the debates over illegal copying sound as if people are attempting to construct a deontological argument somewhat along the following lines:

> The laws established to "protect" the work of wealthy artists and marketed by wealthy and powerful corporations are unjust.
>
> They are unjust because the laws are not the result of a genuinely democratic process, one in which the *consent* of those affected plays the deciding role. Rather, they are laws that result from a legislative process controlled by the powerful – those with the money to do so. Those laws thus represent and protect the interests of the wealthy and powerful – they do not represent or protect the interests of the rest of us.
>
> Given that these laws are unjust, I am allowed (perhaps even obliged) to disobey them.

Perhaps with the help of your instructor and/or cohorts, review some of the important deontological sources for arguments supporting *disobeying* unjust laws (e.g., Ess n.d.).

Can you find/develop deontological arguments along these lines that support *disobeying* prevailing copyright laws as *unjust* laws?

And: if so, how closely do they parallel the sorts of arguments offered by Dr. Martin Luther King, Jr., for example? In particular, how good an *analogy* is there between

the situation and context supporting King's arguments that segregation laws are *unjust* – and thus *must* be disobeyed,

and

the situation and context supporting the arguments you find/develop that copyright laws are *unjust* and thus can or must be disobeyed?

4. Copyright and virtue ethics

Herman Tavani, drawing especially on the work of Michael McFarland (2004), develops a framework for analyzing intellectual property issues that rests squarely on Aristotle's virtue ethics (see ch. 6 for further discussion). On this view, information is taken to have as its ultimate purpose both personal expression and utility; this means in turn that information is best understood as a *common good*, something whose essential nature is to be *shared* – rather than treated, as it is in traditional Western copyright schemes, as an exclusive property. At the extreme, an exclusive focus on information – whether as computer software or a popular song – as a property, the right to which can be controlled by one person or corporation, would lead to the end of "the public domain," i.e., a kind of "information commons" that benefits the whole community. (The analogy here is with the commons in pre-industrial England, i.e., a parcel of land accessible to all for the benefit of all, in contrast with individual and private property.)

Arguably, much good – both individually and communally – has come from the existence of such commons. Indeed, as Niels Øle Finneman points out, part of the Scandinavian approach to information technologies and their supporting infrastructures over the past decades is based on understanding these as common or public goods – ones that thus require and deserve the material support of the state. Direct state support of ICT infrastructure and development has thus contributed to the Scandinavian countries enjoying the highest presence and use of these technologies in their daily lives (Finneman 2005).

From the perspective of virtue ethics, then, we would pursue *excellence* in our abilities to develop, manipulate, and distribute information as a common good – *not* primarily because doing so might benefit us personally in primarily economic terms. Rather, because in doing so (a) we foster and improve upon important capacities and abilities as human beings, including our ability to communicate with one another and benefit one another using these new technologies. As well (b), doing so thereby contributes to greater community harmony and benefit.

To be sure, such an approach, as Tavani emphasizes, is not opposed to individual economic gain. The ideal here would be to develop a system that could conjoin these notions of virtue ethics and the common good with a recognized need for "fair compensation" for the costs and risks individuals and companies take in developing products and making them available in the marketplace. Tavani sees the Creative Commons initiative (discussed above) as one way of institutionalizing such a virtue ethics approach to information (Tavani 2007: 248–55).

Responses? In particular:

A. Are there important *virtues* or habits of excellence that might come into play in either

   (i) practicing *obeying*, e.g., copyright laws (as well as other laws), at least as long as they are *just* laws?
   (ii) practicing *disobeying* such laws?

B. Are there important *virtues* or habits of excellence that might come into play in either

   (i) practicing *obeying*, e.g., copyright laws (as well as other laws) – even if they are *unjust* laws?
   (ii) practicing *disobeying* such laws?

5. Culture – again

(As a reminder: the following generalizations are starting points only – ones that will be accompanied by plenty of counterexamples and that will become much more complex and nuanced as we go along.)

In addition to culture correlating with basic assumptions regarding the individual/community relationship and the nature of property rights (inclusive/exclusive), we have seen that culture may further correlate with the basic ethical frameworks we have been using:

- Roughly, if you have been acculturated in a Western/northern country such as the U.S. and the U.K., it may be that your arguments largely emphasize *utilitarian* approaches.
- If you have been acculturated in a Western/northern country such as the Germanic countries and Scandinavia, it may be that your arguments more likely include *deontological* approaches.
- If you have been acculturated in a non-Western country – especially one shaped by the sorts of traditions we have explored so far (*ubuntu*, Confucian thought, and Buddhist thought) that emphasize the well-being of the community, you may have a stronger likelihood of appreciating *virtue ethics* approaches – i.e., beginning with questions about what kinds of human beings do we need to *become* – and thus what sorts of habits and practices of excellence must we pursue, for the sake of both our own contentment and well-being (*eudaimonia*) and that of our larger community; and/or you may have a stronger likelihood of appreciating the importance of doing what will benefit the larger community in any event, insofar as we as individuals are crucially interdependent with the other members of our community.

What role – if any, so far as you can tell – does your own *culture* play in shaping your attitudes, beliefs, and practices in these matters?

Stated differently: can you see whether or not your own arguments have been reinforced in one or more ways by the larger cultural tradition(s) that have shaped you?

And/or: do your own arguments tend to run *against* the prevailing ethics of the larger cultural traditions that have shaped you?

[After responding to these questions, you may want to revisit the questions regarding our *meta-ethical* frameworks – *ethical relativism, absolutism, and pluralism* – raised above in question (1)(C)(i) and (1)(C)(ii), pp. 88–90].

Additional topics, questions, resources

1. As noted above, there is fierce debate within FLOSS between two approaches – the initial Free Software (FS) movement, affiliated with Richard Stallman and the Free Software Foundation, and the subsequent Open Source Initiative (OSI), begun in 1998 by Eric Raymond and others. Briefly, FS reaches back to a strongly anti-corporate (if not anti-capitalist) "hacker ethic" that emerged in the 1960s. By contrast, OSI intends "to shape free software into an acceptable choice for businesses by defining open source software so that there are no restrictions on distributing it with proprietary software" (Grodzinsky and Wolf 2008: 251). Nonetheless, both approaches share the common goal of seeking to expand the communities who develop and use Free/Libre/Open Source Software (FLOSS).

    These contrasts and debates are of interest in at least three ways. First, especially within discussions of the professional ethics affiliated with computer science, there is significant discussion of the different *ethical* implications of each approach. So, for example, Grodzinsky and Wolf point out that on Stallman's account, FS represents an extension of the Golden Rule, understood in terms of the ability to share freely with others software that we ourselves enjoy and find useful (2008: 247). From this perspective, restricting software (e.g., via copyright) to only those who can pay for it is anti-egalitarian and restricts the benefit of such software to humanity at large (2008: 247). On the other hand, Raymond, relying in part on John Locke's theory of property, argues that the programmer's labor deserves a specific return – primarily, in terms of reputation within the programming community, but certainly not to the exclusion of financial reward (Grodzinsky and Wolf 2008: 259, 262). This means in particular that OSS supports licenses such as the Mozilla Public License (MPL) and Berkeley's Software Distribution (BSD) that allow programmers to modify open source software and then "release it as a proprietary product" (Tavani 2007: 246). These differences, in turn, are important for further considerations of professional accountability and how developers and users – especially within an academic Computer Science department – contribute to the common good (e.g., Grodzinsky and Wolf 2008: 264–9; see also Tavani 2007: 244–7).

Second, the licensing schemes of OSS support a mixed model approach to software development and distribution – exemplified in various commercial distributions of Linux such as Debian, Red Hat, Novell's Suse, and others. While much of the software in these distributions remains free in Stallman's sense – i.e., users can continue to have access to the source code, copy it, modify it, distribute it to others, etc. – some of the software is proprietary; as well, companies are able to turn a profit on these otherwise free distributions by charging for support services, for example. (Indeed, even Microsoft – the pre-eminent exemplar of proprietary software as business – hosts its own Open Source community: Port 25 <http://port25.technet.com/>.) In these ways, OSS softens the otherwise sharp contrast represented by Stallman and FS, on the one hand, and traditional copyright notions, for-profit development of proprietary software, and so forth, on the other hand.

Third, OSS thereby helpfully complicates the correlative contrasts I have initially suggested regarding *culture*. For example, we have explored above a broad contrast between (modern) Western cultures as more oriented towards the individual (and, especially in the case of United States, towards more *utilitarian* ethics and market-friendly approaches – both with regard to privacy and towards intellectual property) vis-à-vis Eastern and indigenous cultures as more oriented towards the community (and thereby, for example, towards *virtue ethics*). To be sure, these cultural contrasts are themselves only initial; they quickly become more complex and nuanced as we learn more about the details of each. So, for example, Soraj Hongladarom's Buddhist account of the person and privacy (ch. 2) highlights Western communitarian traditions (as offsetting Western individualism) and a Buddhist sense of the empirical self (as counterbalancing Eastern emphases on the community.) (See also ch. 2, *Interlude*, pp. 39–43). In addition, OSS further helps to complicate these introductory overviews. For example, its proponents could argue that by taking advantage of (Western) emphases on individual self-interest (as rewarded both in social and financial terms), the licensing schemes of OSS thereby harness the labor and creativity of programmers for the sake of a greater community or public good – namely, the greater (but not entirely free) availability of quality software.

Resources/research

In addition to reviewing available entries, e.g., on Wikipedia, regarding "Open Source Initiative," "Eric Raymond," etc., read into one or more of the following resources (moving from broad introductions to more careful foci):

Tavani, Herman. 2007. Intellectual Property Disputes in Cyberspace (ch. 8), esp. "8.7.2 The Open Source Movement" (pp. 246f.). In *Ethics and Technology: Ethical Issues in an Age of Information and Communication Technology* (2nd ed.), 221–58. Hoboken, N.J.: John Wiley and Sons.
[An excellent overview of IP, including discussion of both FS and OSS.]

Raymond, Eric. 2004. The Cathedral and the Bazaar. In Richard A. Spinello and Herman T. Tavani (eds.), *Readings in Cyberethics* (2nd ed.), 367–96. Boston: Jones and Bartlett.
[Provides a classic account and justification of OSS philosophy and approaches.]

Spinello, Richard A. 2008 Intellectual Property: Legal and Moral Challenges of Online File Sharing. In Kenneth Einar Himma and Herman T. Tavani (eds.), *The Handbook of Information and Computer Ethics*, 553–69. Hoboken, N.J.: John Wiley and Sons.
[A fine overview, focusing on the U.S. context, including attention to the Grokster case.]

Grodzinsky, Frances S. and Wolf, Marty. 2008. Ethical Interest in Free and Open Source Software. In Kenneth Einar Himma and Herman T. Tavani (eds.), *The Handbook of Information and Computer Ethics*, 245–71. Hoboken, N.J.: John Wiley and Sons.
[An extensive analysis of the important philosophical differences between FS and OSS.]

Gordon, Wendy J. 2008. Moral Philosophy, Information Technology, and Copyright: The Grokster Case. In Jeroen van den Hoven and John Weckert (eds.), *Information Technology and Moral Philosophy*, 270–300. Cambridge: Cambridge University Press.

[Analyzes this famous ruling from both consequentialist and deontological perspectives. Gordon finds in particular that a Lockean theory of property fails to support the Court's ruling against Grokster as intending to foster copyright violation.]

Moore, Adam D. 2008. Personality-Based, Rule-Utilitarian, and Lockean Justifications of Intellectual Property. In Kenneth Einar Himma and Herman T. Tavani (eds.), *The Handbook of Information and Computer Ethics*, 105–30. Hoboken, N.J.: John Wiley and Sons.
[Moore argues that the Lockean justification emerges as the strongest of the three.]

Chopra, Samir and Scott Dexter. 2008. *Decoding Liberation: The Promises of Free and Open Source Software*. New York: Routledge.
[Chapter 1, "Free Software and Political Economy," provides an extensive analysis of the contrasts between FS and OSS, including a Marxian approach. Available online: <http://port25.technet.com/videos/research/chapter1_chopradexter.pdf>, accessed September 15, 2008. *Review*: Hill, Benjamin Mako. *Minds & Machines 18* (2008): 297–2. Available online: <http://www.sci.brooklyn.cuny.edu/~schopra/MMDL.pdf>, accessed September 15, 2008. Hill argues that the philosophical differences between FS and OSS emphasized by Chopra and Dexter are largely irrelevant in the *praxis* of FOSS – that individuals and groups from both "sides" work together, if not always harmoniously, and that these philosophical differences "might be better explained as tactical, social, or political differences" (p. 299).]

A. Given the resources you have examined, highlight what seem to be the most important *philosophical* differences between FS and OSS regarding intellectual property and how it is to be fairly and justly developed and shared.

   (i) Given these differences, which of the two approaches seems more defensible, in your view?

   (ii) In light of the differences, which approach seems closer to your own actual practices and behaviors regarding, e.g., music and other forms of file-sharing?

B. Review your responses to question 1, above (pp. 87–90). In light of the contrasts we have now seen between FS and OSS, we can subdivide the "FLOSS/copyleft" position on the continuum diagram in turn:

US/Europe	FLOSS/copyleft	Confucian
	OSS ◄────► FS	*Ubuntu*

◄──────────────────────────────────────►

Greater stress on	Greater stress on
individual	*community*
exclusive property rights	*inclusive property rights*

Given this more complex range of choices regarding possible philosophical positions on intellectual property, would you change any of your responses to the various questions posed in (1), especially (1)(B) (p. 88 on possible correlations between your cultural background(s) and your ethics)? If so, in what ways? If not, why not?

Alternatively: are the philosophical differences between FS and OSS in fact relevant to your own cultural background and ethics? And/or: are these differences, as Benjamin Hill suggests in his review, less about philosophy and ethics and more about tactics, etc.?

2. A *universal* ethics of copying? [*Meta-ethical* issues vis-à-vis the cultural backgrounds of norms and practices of copying.]

In light of these large contrasts between how different cultures approach the ethics of copying, can there be/ought there to be a universally shared ethics of copying and thereby a universal set of laws (copyright/copyleft/other?) governing our copying and distributing the works of others?

It may be useful to review the discussion of ethical pluralism in chapter 6, along with the ethical pluralism described in chapter 2 regarding a global ethics of privacy and data privacy protection.

In addition, you might want to look at

Brey, Philip. 2007. Is Information Ethics Culture-Relative? *Journal of Technology and Human Interaction* 3 (3), 12–24.

Brey examines the arguments for *ethical relativism* in the face of the sorts of cultural differences we have examined here – and with explicit attention to Confucian, Buddhist, and African thought with regard to both privacy and intellectual property rights. As with several other philosophers in Information Ethics – or what Rafael Capurro rightly calls Intercultural Information Ethics (2005) – Brey argues for an *ethical pluralism* as a middle ground between *ethical relativism* and *ethical absolutism.* (Again, see ch. 6 for a fuller discussion of these meta-ethical frameworks.)

Additional readings, suggested resources for review

University of Minnesota Libraries n.d. Copyright Information and Education: Working with Fair Use. <http://www.lib.umn.edu/copyright/ >, accessed September 15, 2008.
　　[Includes useful background information on U.S. copyright and fair use, including a "Copyright Decision Map" and "Fair Use Analysis Tool" designed to help students and faculty work through four factors (purpose, nature, amount, market effect) weighing in favor of/against "fair use" of copyrighted materials.]

"P2P or File Sharing" (Educause) <http://connect.educause.edu/term_view/P2P%2BFile%2BSharing>, accessed September 15, 2008.
　　[A good (but exclusively U.S.-based) collection of articles, rulings, campus guidelines, etc. on file-sharing.]

"Piracy: Online and On The Street" (Recording Industry Association of America) <http://www.riaa.com/physicalpiracy.php>, accessed September 15, 2008.
　　[Current initiatives, related articles, and rationale for the RIAA's anti-piracy campaigns.]

Himma, Kenneth Einar. 2008. The Justification of Intellectual Property: Contemporary Philosophical Disputes. *Journal of the American Society for Information Sciences and Technology 59* (7: May): 1143–61.
　　[Provides an exceptionally extensive and thorough discussion of major Western positions, their sources and supporting arguments – arguing, in the end, for at least limited rights over content creation and control.]

Snapper, John. 2008. The Matter of Plagiarism: What, Why, and If. In Kenneth Einar Himma and Herman T. Tavani (eds.), *The Handbook of Information and Computer Ethics,* 533–52. Hoboken, N.J.: John Wiley and Sons.
　　[As the title suggests, Snapper is primarily concerned with what counts as plagiarism in the digital age, but his discussion includes an excellent overview of the larger frameworks for thinking through

matters of intellectual property and copyright within a U.S. perspective, but with some attention to the French tradition as well.]

Resources on FLOSS/ICT4D (ICT for Development)

UNDP–APDIP International Open Source Network. <http://www.iosn. net/>, accessed September 15, 2008.
[International Open Source Network – an initiative of the United Nations Development Programme's (UNDP) Asia Pacific Development Information Programme (APDIP) and supported by the International Development Research Centre (IDRC) of Canada.]

Asia-Pacific Development Information Programme (APDIP). n.d. Case Studies on Free and Open Source Software. <http://www.apdip.net/resources/case/foss/>, accessed January 20, 2008.

FUNREDES. n.d. *Fundación-Redes-y-Desarrollo/*Networks-and-Development-Foundation/*Association-Réseaux-et-Développement*. <http://www.funredes.org/english/index.php3>, accessed January 20, 2008.

Pimienta, Daniel. 2007. *Digital Divide, Social Divide, Paradigmatic Divide*. <http://funredes.org/mistica/english/cyberlibrary/thematic/Paradig matic_Divide.pdf>, accessed January 20, 2008.

Servaes, Jan (ed.). 2007. *Communication for Development and Social Change*. Los Angeles: Sage.

CHAPTER 4

Citizenship in the Global Metropolis

Cultures do not talk to each other; individuals do.

(Scollon and Wong-Scollon 2001: 138, cited in Hewling 2005)

Chapter overview

We begin with a case-study that highlights how the global scope of digital communication media radically amplifies the potential consequences of our communication, and thereby makes the ethics of cross-cultural communication more and more central for all users of digital communication media. At least initially, these ethics are shaped by our cross-cultural encounters as *embodied* beings. That is, our ethical responses to one another are often matters we feel through "in our heart" or by "what my gut tells me," not just think through in a purely conceptual and abstract way. Moreover, we communicate a great deal to one another *with* our bodies (gesture, gaze, body distance, etc.) what we feel and how we respond to one another, most especially when it comes to important ethical issues and differences. This recognition of the roles of the body in our cross-cultural encounters with one another then emphasizes the importance of a range of *virtues* (the habits, practices, and qualities that contribute to our becoming more complete and content human beings) required for such cross-cultural communication. These virtues are especially important if we are to move from communication that focuses solely on our shared commonalities and identities to communication that further allows us to develop deeper understanding of and respect for the (sometimes radical) differences that shape our individual and cultural identities. Coming to understand these differences is crucial not only to becoming more complete and content human beings; understanding these differences is necessary if we are to avoid a naïve ethnocentrism, the belief that our own cultural norms, beliefs, and practices are somehow universal. Such ethnocentrism is further to be avoided because it runs the risk of fostering cultural imperialism – our imposing our own cultural norms, beliefs, and practices on others.

We then consider how online environments, while radically increasing our opportunities for cross-cultural communication, also increase the risks of ethnocentrism and cultural imperialism. Briefly, our encounters with one another online as disembodied beings thereby leave out many (if not most) of the signs of our differences with one another – signs that are unmistakable in embodied encounters. Moreover, we will see how our communication technologies themselves embed and foster the specific cultural values and communicative preferences of their designers – such that their diffusion in "target" cultures risks imposing these values and preferences. In both ways, as online environments *increase* the possibilities of communication that risks becoming imperialistic, they thereby make our reflection on the ethics of online communication that much more urgent.

Case-study: the Muhammad cartoons

On September 30, 2006, a conservative Danish newspaper, *Jyllands-Posten*, published 12 cartoons, subsequently known as "the Muhammad cartoons." The cartoons invoked a number of Western stereotypes about the Prophet Muhammad and Islam more generally, and were intended to be provocative, if not offensive. According to the editor who commissioned the cartoons, Flemming Rose, publishing the cartoons was in part a protest against what he perceived to be an increasing self-censorship in Europe in publishing on Islam and Muslims in Europe (Rose 2006, cited in Debatin 2007: 13).

The publication of the cartoons resulted in far more than the hoped-for debate regarding freedom of speech and democracy vis-à-vis any special claims for restricting public expression in the name of avoiding offense to religious sensibilities. Both peaceful and violent protests followed, beginning in Denmark but then spreading quickly throughout the world. While some newspapers demonstrated solidarity with *Jyllands-Posten* in particular, and freedom of the press more generally, by republishing the cartoons, the Council of Europe condemned the Danish government for failing to do anything about the cartoons. The next six months were marked by extensive protests, eventually resulting in scores of deaths, alongside diplomatic objections in the form of ambassadors being withdrawn. But while *Jyllands-Posten*, as demanded by Palestinian gunmen following a raid on the European Union offices in Gaza, apologized for the cartoons on its website, both German and Danish courts ruled that publishing the cartoons was protected by national law. (See Debatin 2007: 228–32 for a more extensive timeline.)

Questions for reflection, discussion

1. If you had been the chief editor of *Jyllands-Posten*, would you have approved of the publication of the Muhammad cartoons?

    As always, the chief question for our purposes is, *why?* That is, whatever your response to this question, what *reasons, arguments, grounds, feelings, intuitions, sensibilities*, etc., support and provide justification for your position?

2. *Perspective-taking*: take an alternative view/judgment to the one you articulated and defended in question 1. What sorts of *reasons, arguments, grounds, feelings, intuitions, sensibilities*, etc. can you imagine might support and provide justification for this alternative position?

3. The Muhammad cartoons have become an iconic example of how contemporary digital media increasingly weave together countries and peoples in ways undreamed of even a few decades ago – and not always to beneficent effect. Contrary to Marshall McLuhan's vision of an increasingly democratic and peaceful "electronic global village," the Muhammed cartoons rather illustrate the power of contemporary information and communication technologies (ICTs) and computer-mediated communication (CMC) to fan the flames of extant political, religious, and social differences, not resolve them. That is, had *Jyllands-Posten* published these cartoons in the pre-Internet era, it's unlikely that anyone outside of Denmark would have known about it – much less, have been in a position to turn it into the occasion of violent protest and diplomatic wrangling.

Either individually and/or in collaboration with either a small group or your class at large, consider the arguments for and against publishing the cartoons. Do any of these arguments depend on the impacts and consequences of our actions, as these are now potentially distributable more or less instantaneously throughout a global range of cultures? If so, which ones?

To put it differently, does the emergence of digital media – in this case, especially the Internet and the Web – significantly alter how we are to think about our ethical responsibilities towards one another? If so, in what ways? If not, why not?

Cross-cultural communication online: initial considerations

As especially the Internet and the Web dramatically increase our abilities to communicate across a global range of cultures, they thereby introduce a range of correlative ethical difficulties – some of which, as affiliated with cross-cultural encounters, have been with us as long as human beings have lived together in culturally and linguistically distinct groups. Others may be distinctively new – or, at least, dramatically amplified versions of older ones.

At a first level – and as suggested by the example of the Muhammad cartoons – a cluster of these difficulties are fairly obvious and largely represent old problems now elevated to an electronic level. So, for example, an advertising agency may be called in to develop an advertising campaign to attract Muslim tourists to Malaysia. Both respect for others and the simple interest in not offending our potential customers tell us that our images – especially our images of women – should follow the guidelines of the religion shaping the values and sensibilities of those customers. Hence, we would generally want to avoid images of women that violate standard dress code for Islamic women – i.e., anything that would show bare arms and/or shoulders, and/or an uncovered head.

It turns out, however, that even at this first level, things are not so simple. While these rules would strictly apply in some countries (e.g., Saudi Arabia), they do not hold in Malaysia, as an officially Islamic country that nonetheless incorporates a wide range of religious traditions, including Buddhists, Christians, Hindus, and Confucians (Hashim et al. 2007). The result is a considerable tolerance and understanding of "Other" peoples and their practices, and, in particular, the use of the image of a Kadazan woman (one of the ethnic groups making up Malaysia's population) showing uncovered arms, neck, and head is in fact an acceptable practice (see, e.g., <http://www.sabahtourism.com/culture.php?ID=14>, accessed December 11, 2007).

Nor is such an image the only exception. Noor Hazarina Hashim writes:

> The Kadazan traditional costume is just one of the examples [of images possibly questionable for Islamic viewers].
>
> We have the Chinese wearing Cheongsam and the Indian wearing Saree, all of which would fall under "scantily clad women" . . . according to Islamic dress code. However, most Malaysians tend to be less sensitive and accept that as a part of Malaysia melting pot – something that "belongs" to Malaysia. (Email, January 27, 2007)

Hence, as anyone knows who has tried to be a respectful guest in an "Other" culture, it is not always easy to know what to do, even if you believe yourself to be well informed about important rules, values, practices, etc. While these general rules and characterizations about a culture, religion, etc., are a critical starting point, they are only a starting point. That is – again, as anybody knows who has attempted to visit an "Other" culture for an extended period of time (much less, tried to fit in as an immigrant) – much depends upon the fine-grained details of a local/specific context. These details only begin with very broad generalizations about cultures, religions, etc., and quickly become very complex, even contradictory, as we move into the everyday, "street" level of a given group of people, living in a given community in a given city or rural area in a given part of a country. To make things even more complicated, the cultures that influence and shape our behaviors and beliefs – including our basic values and communicative preferences – include additional layers: for example, depending on our age and gender, we may participate in and/or be influenced more or less by a local/regional/national/international "youth culture." (Cf. ch. 2, *Interlude*, pp. 39–43.)

Finally, as the Muhammad cartoons illustrate, there are often deep – perhaps irresoluble – conflicts between diverse sets of basic values, norms, practices, beliefs, etc., as these define in large measure both individual and collective identities for people, groups, and whole societies. That is, what is seen as a thoroughly justified expression of a modern/Western insistence on the untrammeled freedom of speech and expression is seen in this instance – also with considerable justification – as an inexcusable exercise in blaspheming the holy Prophet of Islam. For believers, especially in those countries and traditions *not* deeply shaped by modern Western Enlightenment ideas of the separation of religion from the (secular) state, such images can only be interpreted as yet one more expression of contempt and misunderstanding – an expression that only pours more salt into the deep wounds of centuries of Western colonialism and exploitation.

How, then, might we communicate effectively across such deep cultural differences? More precisely, what *ethical guidelines* might there be for effective cross-cultural communication online?

Ethics for cosmopolitans (citizens of the world) – offline and online: from consequentialism to virtue ethics

A first component of an ethics for cross-cultural communication online is the obligation to become familiar with the cultures of the world, beginning with the cultures of those with whom one frequently communicates.

Such an obligation is obvious in the offline world: it has been a component of Western liberal arts education since the Renaissance – and traces back to the earliest Greek encounters with radical cultural diversity. This obligation is obvious if only for strictly self-interested and *consequentialist* reasons: most basically, any group of people who have sought to trade (relatively) peacefully with another group, as characterized by sometimes radically different beliefs, values, practices, etc., has learned that at least some knowledge of "the Other"[1] is needed to avoid potentially fatal misunderstandings.

But our interest in profiting from our encounters with one another is a limited motivation for learning more about the Other. That is, such an interest may limit us to focusing primarily on what we share in common – beginning with, for example, a *lingua franca* that allows us to barter, exchange, and make contracts. To be sure, such interests and motivations will carry us far, and are sufficient for many intents and purposes. But this focus on commonalities further carries a specific risk – one that can be fatal to effective cross-cultural communication, especially in online contexts. As long as I focus only on our shared commonalities (and especially if I only see the Other as a "target" such as a resource or market that I seek to exploit for my own advantage), I never move to a second stage of human communication and understanding: I never move, that is, to an appreciation of "the Other as Other" – the Other as *irreducibly different* from me. Yet, as innumerable philosophers, religious teachers, and humanists have stressed, it is only when I encounter the Other as Other, *including* a fundamental respect for and acceptance of the sometimes radically different norms, beliefs, and practices of the culture(s) that shape the Other, that I move to a more complete understanding of the Other.

This understanding may have no marketable value whatsoever. But as anyone who has developed a deep friendship with someone from a different culture knows, such friendship and understanding of the Other as Other carries with it two ethically crucial elements. First, we thereby learn to accept and respect Others not simply as similar to ourselves, but also as (sometimes radically) irreducibly different from ourselves. This deep understanding of

---

1   In this context, I use "other" – i.e., without a capital – to signal a viewpoint or perspective on the "other" as different and thereby, at least initially, as suspect, potentially dangerous, inferior, etc. (This is the viewpoint characteristic of ethnocentrism and related perspectives of racism, sexism, etc.) By contrast, as the phrase "Qther as Other" suggests, Other with a capital is intended to convey that we recognize the Other as fully equal, fully human, while simultaneously irreducibly different from us.

   This convention echoes Levinas's usage in his analysis of "the Other as Other," as a positive "alterity" (e.g., Levinas 1987).

"the Otherness of the Other," second, provides a perspective or standpoint for looking back on our own culturally specific norms, beliefs, practices, etc. The Western tradition has stressed that such a perspective is essential, not only for the sake of overcoming ethnocentrism and its attendant dangers, but also if we are to move from simple acceptance of the norms, beliefs, practices, etc., of the larger culture around us to a more *autonomous*, informed *choice* regarding what we ourselves believe is good, true, and beautiful.

In both ways, as those who have these experiences know, we are immeasurably enriched, if not in material terms, then in human terms. To begin with, as we expand our perspectives in these respects, we come to a profound new understanding of ourselves, and are thereby able to *construct* ourselves – out of new choices made possible only in such deep encounters with the Other – in ways that would otherwise be impossible. At the same time, this deep self-enrichment is made possible only by our taking up an attitude or posture towards *difference*: we learn to shift from the initial uncertainty, fear, and hostility that deep cultural differences often inspire, to a posture that is more open to and actively interested in such differences, both as they define the Other as Other. Particularly from the standpoint of virtue ethics, learning to encounter and respond to difference in these new ways is especially important: we thereby acquire (in part, as we learn to practice and improve upon) our abilities to understand and respect these differences such that we may recognize and insist upon the human dignity of the Other as Other – not simply as the Other is like us. And, as we will see, doing so entails the practice of a specific set of basic virtues – qualities and abilities that we acquire and improve upon over time through practice: these include humility, understanding, compassion, and forgiveness.

From the standpoint of this more comprehensive experience and perspective, we can then more clearly see a further problem that follows from our initial interests in the Other as target and resource, and our correlative focus on the commonalities that allow us to communicate and trade with one another. The focus on commonalities prevents what is sometimes called "culture shock" or "ontological shock" – i.e., precisely the recognition that the way

Others believe, practice, and live are radically different, and, thereby, that *our own norms, beliefs, and ways of life* are *not* universal. For the most part, if we are *not* shocked out of the ethnocentric belief that "our ways are everyone's ways," then it makes it much easier for us to proceed from ethnocentrism to imperialism, i.e., the imposition by force (whether subtle or gross) of our ways upon others.

Hence the stress in the Western tradition (and, for example, Confucian thought) on *travel* as an essential component to becoming more fully human. As we travel and encounter Others in sometimes strongly different linguistic and cultural contexts, we are forced to recognize that our ways are not everyone's ways.

A key virtue that both fosters and is fostered by these more radical encounters with the Other as Other is *epistemological humility*. That is, especially once we have experienced such cultural and ontological shocks, once we are convinced that our ways are not necessarily the ways of everyone else, we are more inclined to practice a humility regarding our own language, beliefs, norms, and practices. That is, we are more likely to *know* at a deep, experiential level that these basic cultural elements are *limited* and not universally shared. Hence, we cannot (naïvely) assume that our ways "work" for everyone else.

By the same token, when Others recognize in us this sort of humility – in contrast with a naïve ethnocentrism that runs the risk of imperialism – they are far more likely to practice the virtues of patience, understanding, and forgiveness when we do make the mistakes inevitable in crossing cultures. As we will see in chapter 6, these virtues are not only stressed within traditions of virtue ethics as such (in the West, beginning with Socrates and Aristotle, and stretching into contemporary expressions, and in the East, e.g., in Confucian thought). In addition, they are endorsed in the major religious and philosophical traditions of the world, including the Abrahamic religions (Judaism, Christianity, Islam) and Eastern traditions such as the diverse forms of Hinduism and Buddhism. Such virtues, as shared among the diverse cultures and traditions of the world, may thus serve as crucial "bridge" values in a genuinely *global* ethics – one that enjoys validity and legitimacy among most (if not all) cultures, not simply a particular one.

However that may be, here we can see how these virtues are essential to cross-cultural communication. Again, errors – sometimes serious errors – are inevitable, especially as we first begin our efforts to understand and learn how to negotiate the sometimes radical differences between our culture and those of the Others with whom we wish to communicate. But especially as others see that we are seeking to practice humility in our efforts, e.g., to learn how to speak and behave properly in a new culture – and thereby, that we are *not* naïvely and ethnocentrically assuming that our ways are everyone else's ways – they are far more likely to recognize our errors for what they are: errors committed out of ignorance, not arrogance – and certainly not the arrogance of an ethnocentrism that risks becoming an imperialism that seeks only to reduce the Other to an exploitable sameness.

In sum: our experiences with cross-cultural communication in the "offline" world argue that if we seek to know the Other as Other – i.e., in ways that respect and preserve the irreducible differences that define our distinct individual and cultural identities – then we are obliged to practice the virtues of humility, understanding, compassion, and forgiveness.

Perhaps paradoxically, however: while online venues dramatically expand our opportunities for cross-cultural communication, such venues present two significant obstacles to our acquiring and practicing the virtues needed for cross-cultural communication that includes fundamental respect for the Other as Other. To begin with, as online venues make possible largely *disembodied* encounters, they thereby sidestep our encounters with one another as embodied beings, where our embodiment carries the multiple and undeniable signs of our radical differences from one another. Online venues thereby help us avoid the sorts of culture shocks and ontological shocks that follow upon our encounters with one another as embodied beings – and thereby make it easy to ignore differences. Doing so, finally, opens the door (again) to focusing on commonalities and thus raises the risks of ethnocentrism and imperialism. Secondly – and more subtly – we will see that the technologies of online communication themselves embed and foster the specific cultural values and communicative preferences of their designers. Thereby, their very use in "target" cultures again runs the risk of a cultural imperialism that imposes the beliefs, norms, and practices of one culture upon another. (In exploring this second risk, we will also expand our understanding of diverse cultures – thereby, I hope, helping us to begin to explore cultural differences in concrete ways, ways that may further help us overcome ethnocentrism and avoid imperialism.)

The risk of disembodiment

In the early days of the Internet and the World Wide Web, there was much enthusiasm for "life online" as a *disembodied* life in virtual worlds. In particular, there were hopes that by stripping away the various elements of an *embodied* self – including our identifying markers in terms of gender, members of specific groups and cultures, etc. – online communication might lead to a greater gender and social equality in virtual worlds. But while disembodiment might lead to ethically desirable (if not politically urgent) ends, it soon became apparent, however, that disembodiment also led to darker communicative possibilities as well. As but one example: our disembodied anonymity online appears to foster such well-known phenomena as *flaming* and other forms of aggressive speech. Since our own embodied selves are hidden – and since we do not experience the Other in close, physical proximity to ourselves – it is easier to engage in aggressive speech: we do not directly experience the impacts of this speech on others ("it's only words"), nor can we be easily found out as physical targets for the sorts of verbal (and perhaps physical) retribution that such aggressive speech would engender in face-to-face encounters.

By the same token, when the Other is re-presented to us online as a disembodied voice – minimally, a text written in our own language – then we experience none of the radical Otherness of the Other that powerfully confronts us in our real-world encounters. That is, when we encounter one another as embodied beings face to face, our full differences from one another are, so to speak, on display. These begin with our language and accent, but further include our dress, voice, gesture, actions – the whole range of postures and behaviors that define our sense of who we are both as distinctive individuals *and* as members of distinct cultures. It is as we experience these first-hand and in close proximity to one another that the full *difference* of the Other strikes us and cannot be ignored. In other terms: our disembodied encounters with one another, as they focus on the commonalities needed for communication, thereby run the risk of failing to give us the ontological/culture shocks that force us to recognize the multiple ways in which much of our own beliefs, norms, and practices are limited to our own culture. Rather, as many – if not most – of these indicators of (sometimes radical) difference are absent in online venues, so the constant and powerful awareness of radical difference is absent. And in its absence, it is correspondingly easier to presume that "the Other" is indeed just like us. Hence it is correspondingly easier to fall back into our own ethnocentrisms. And instead of being constantly reminded of the need to exercise epistemological humility and to avoid the dangers of ethnocentrism leading to cultural imperialism, it is rather easy to impose uncritically our ways on others, on the assumption that they are just like us. This is to say: the absence of the embodied self and its multiple cues of radical difference in online contexts make it easier for us to behave in an ethnocentric and thereby imperialistic fashion.

Hence, online venues thereby *increase* the importance of our exploring in depth the multiple cultures of Others online – at least, if we seek to know the Other as Other, not solely as an exploitable target and resource. This would seem to mean, then, that insofar as digital communication media make us *cosmopolitans* (citizens of the world) by default, we then face an *increased* obligation to acquire and practice the virtues of humility, understanding, compassion, and forgiveness, insofar as these virtues are necessary conditions to a cross-cultural communication that allows us to know one another not only as the same but also as radically and irreducibly different from ourselves.

This much may seem relatively obvious – even trite. But there is a further dimension of online communication that, as more subtle, commands still more careful attention.

Cross-cultural communication online: when the technological solutions become part of the problem . . .

A second obstacle to cross-cultural communication that respects and fosters radical differences is presented by the communication media themselves. That is, we can now explore with some care how these media, rather than serving simply as neutral tools, in fact embed and foster specific cultural values and communicative preferences. Doing so will help us begin to learn more about important cultural differences, and thereby contribute, I hope, to our becoming more adept as communicators able to recognize, respect, and foster essential differences. At the same time, understanding how our tools of communication themselves may serve as agents of cultural imperialism will help us better avoid such imperialism.

There is a strong tendency to assume that the technologies of CMC – the computers, networks, and various software applications that make it possible for us to use them to communicate globally – are somehow "just tools," i.e., somehow *neutral* media that transparently and without particular bias allow us to communicate with one another. But in fact, a now considerable body of research shows just the contrary: ICTs, as designed by particular people from a given culture, thereby reflect the cultural values and communicative preferences of their designers. Such communication tools thereby foster and reinforce the values and preferences of their designers – while directly clashing with the values and preferences of those in "other" cultures. What this means is, first of all, that the diffusion of Western-designed ICTs throughout the globe threatens to result in a kind of computer-mediated cultural imperialism – i.e.,

> a *reshaping* of especially the language(s), fundamental values, communicative preferences, and practices that are characteristic of a given culture, as these work to shape and influence both individual and group *identity, without* the "informed consent" of the participants in the "target" cultures.

That is, as Western-designed ICTs are introduced in and used by people in diverse cultural settings, those people often have to adapt their prior ways of doing things – first of all, in terms of the *communicative preferences* characteristic of their culture, in order to make effective use of these technologies, whose design favors sometimes very different communicative preferences.

High-context/low-content vs. low-context/high-content

So, for example, Edward T. Hall (1976) has developed an important – and now widely recognized and applied – distinction between *high-context/low-content* cultures (exemplified in such countries as Japan, China, Thailand; many countries in the Middle East; and many indigenous peoples throughout the world) and *low-context/ high-content* cultures (such as the United States, the United Kingdom, and the Germanic countries, including Scandinavia). As the name suggests, low-context/high-content (LC) communication styles emphasize the transmission of information with relatively little attention to the larger, especially *social*, contexts. This social context is shaped by such factors as: individual status vis-à-vis those with whom one is communicating; and social networks that range from informal friendship networks to professional networks, family and kinship relationships that help define one's social role(s) and status in the larger community, etc. These sorts of elements, by contrast, are emphasized in high-context/low-content cultures (HC). While LC cultures favor direct and explicit communication, HC cultures prefer indirect forms of communication – including a greater emphasis on non-verbal elements such as body distance, gesture, and gaze, as these signal, for example, whether one is communicating with a superior (and thus likely required to lower one's eyes) or an equal (in which case, direct gaze may be more appropriate – rather than rude). Advertisers have learned to skillfully exploit these differences, finding that their advertising is more effective as it is "tuned" to the communicative preferences of their target audience. So Marc Hermeking (2005), summarizing over ten years of research in this field, notes that HC cultures favor indirect messages that create emotions, especially through pictures and entertainment elements, while LC cultures favor more direct messages that convey more information about the product.

Not surprisingly, ICTs, as initially designed in Western countries – primarily the United States – favoring an LC communication style, are clearly oriented towards facilitating LC forms of communication. For example, a basic (text-only) email can contain a great deal of *information*, but usually reveals little, if anything, about *context*, including the social relationships and relative status of the sender and receiver. By contrast, as Lorna Heaton (2001) has documented, a Computer-Supported Cooperative Work (CSCW) system designed by and for Japanese facilitated the kinds of communication required in Japan as an HC society: that is, using a video system that allowed collaborators to see one another at their work, the system thereby allowed them to communicate in non-verbal terms – gesture, gaze, and body distance.

The point of these examples is to show that our technologies are never "just tools," somehow culturally (and/or communicatively) neutral. Rather, technologies – most certainly those designed for the sake of communication – embed and favor a specific set of communicative preferences (usually those of their designers). (For further reading and research resources, see *Additional research/ writing, 2. Cross-cultural communication – offline/online* (below, pp. 129–30), and *For further reading, on cross-cultural communication: Hall, Hofstede, and "the Other" online* (below, pp. 131–2).)

There is nothing necessarily wrong with this. The danger comes, however, when

(a) we – designers and users – *assume* that *our* cultural values and communicative preferences are genuinely "universal" (when they're not), and thereby

(b) further assume that the technologies we design and use – including our communication technologies – are indeed neutral in terms of the communicative preferences and cultural values that they reflect and foster, so that

(c) we assume that we can diffuse these technologies in any cultural setting whatsoever without harm.

On the contrary, as I hope this first contrast between HC and LC communication styles helps to make clear, diffusing a given communication technology, as it embeds and fosters the specific communicative preferences of its designers, thereby runs the risk of – however inadvertently or unintentionally – thereby imposing those communicative preferences on people in "target" cultures who may thereby be immersed in and deeply shaped by very different communicative preferences.

In other words, the 1990s' enthusiasm to "wire the world" – ostensibly in the name of universal values such as emancipation, freedom of expression, greater democracy, etc. – as it thus encouraged the diffusion of ICTs designed around and fostering LC forms of communication, thereby (however inadvertently) worked to impose – usually without discussion, explicit awareness, much less consent – the communication styles of one cultural group upon another (Ess 1998).

Individualism and freedom of expression

The same point can be made in terms of basic *values* which help shape our identities as members of a given (set of) culture(s).

Just as ICTs embed and foster specific communication styles, so their design and use reflects at least some of the basic values and beliefs of their designers. As a primary example, in the early years of the Internet and CMC, both technologies were hailed in the U.S. especially as fostering *individual freedom of expression* and thereby greater *democracy*. "The Internet interprets censorship as damage," the saying went, and the possibility of *anonymous* communication online fostered hopes that, for example, subordinates in a company might feel freer to email the boss with a complaint or suggestion, thereby flattening hierarchies.

This is all well and good, but what about countries and cultures that do not share these distinctively Western values? What happens when ICTs – designed precisely to foster individual communication ("personal" computer) and freer communication – are introduced into cultures that emphasize more strongly the importance of the community or larger *collective*, and thereby the importance of hierarchies, as such hierarchies mean not simply subordinates' dutiful obedience to superiors, but also the obligation of superiors to tend to the needs and interests of their subordinates?

Not surprisingly, there are often significant cultural clashes and "failures to communicate." So, as but one example, Sjarif Abdat and Graham P. Pervan (2000) have documented how a Group Support System (GSS) used in Indonesia led to major problems – precisely because of its intended feature of anonymous communications. In terms of the cultural axis introduced by Geert Hofstede (1980, 1983, 1984, 1991), while the United States may be characterized as a *high-individualism/low-collectivism culture*, Indonesia, in part as influenced by Confucian tradition, is a *low-individualism/ high-collectivism* culture. While acknowledging that anonymous communications might help with "pre-meeting" discussions among sub-committees, etc., insofar as anonymity encouraged questions of superiors, such questions were (properly) interpreted as threatening "face" and the authority of such superiors. In light of this profound clash between the cultural values embedded and fostered by the GSS and those of the "target" culture, Abdat and Pervan argue that anonymity should be treated as a switchable feature – i.e., something that can be turned off in order to make the technology more culturally appropriate (2000: 213f.).

Again, the point here is that ICTs – most especially those designed to help us communicate with one another – are *not* somehow "neutral" or "just tools"; rather, they embed and foster specific cultural values and communicative preferences. In doing so, as we take up and diffuse these technologies globally, again, we thereby run the risk of imposing upon others a suite of cultural values and communicative preferences quite different from their own. In contrast, say, with gun boats, genocide, and other forms of enforced colonization and imperialism, this sort of imposition can be all the more powerful and insidious precisely because it runs below the surface of our conscious intentions and awareness. (To use an extreme analogy: it is simple to give blankets to Native Americans, thinking that one is doing good. And for the recipients, there might not be anything obviously dangerous about the blankets, so that their accepting them looks like free and informed consent. But if the blankets carry smallpox, the giver contributes to genocide nonetheless.)

As I hope is clear by now, it's the *covert* process at work here that makes the consequences of such exchanges both very effective – it's hard to defend yourself against a cultural change that may be masked by an apparently free choice of accepting a potentially beneficial technology – and ethically troubling.

Discussion/reflection/writing

1. So what's wrong (if anything) with cultural imperialism?

   Cultural hybridization – a sort of mixing up of artifacts, behaviors, values, etc., from diverse cultures, resulting in a new "third culture" or identity, is perhaps as old as human cultures themselves. Such hybridization – represented, for example, by the incorporation of "pagan" customs and figures into Christian celebrations of Christmas – is perhaps less ethically problematic because it appears to rest upon *choice*: that is, it might be argued that such hybridization is ethically justified and acceptable because people are free to *choose* whether or not to take up the artifacts, practices, etc., of an "other" culture and make it part of their own daily life.

   Cultural imperialism, however, represents something different – a transformation of values, preferences, practices, etc., *forced* (either *overtly* and/or *covertly*) upon "target" cultures, with the clear intention of exploiting the target cultures for the benefit of the dominant culture(s). Such imperialism, indeed, can be fatal both for individuals and for cultures (defined in this context as social groups whose patterns of behavior, values, traditions, etc., persist through several generations). Hence there are two basic objections to cultural imperialism: one, it denies freedom of choice; and, two, it violates what is argued to be a *right* to individual and cultural identity (so the United Nations' Universal Declaration of Human Rights, 1948).

   Review the discussion above regarding especially *covert* forms of cultural imperialism – specifically, the claim that the diffusion throughout the globe of ICTs and forms of CMC originating in Western cultures may contribute to a kind of covert cultural imperialism as this diffusion may force the acceptance of cultural values (e.g., *individualism*) and/or communicative preferences (e.g., low-context/high-content).

A. Insofar as you find this discussion persuasive, is the resulting cultural imperialism indeed ethically objectionable? If so, why – i.e., on what grounds? If not, why not – and again, on what grounds?

B. Who, if anyone, is *responsible* for the impacts and consequences of the global diffusion of ICTs – especially when those impacts and consequences are ethically problematic? That is:

- the designers of the hardware? designers of the software?
- the companies that build and distribute these technologies?
- the users – either within the developed world or in the developing world?
- some other actors (please specify)?

If you think that ethical responsibility can*not* be restricted to just one person or institution, you may be moving to a notion of *distributed responsibility*. This will be a topic for more advanced research and writing.

2. [*Consequentialism.*] A consequentialist approach to ethical problems seeks to undertake a kind of ethical cost–benefit analysis of possible choices, weigh these against one another, and then determine the better choice based on the ostensibly most beneficial outcome for a specific group ("the greatest good for the greatest number").

A. How might a consequentialist begin to analyze the following:

- On the one hand, the global diffusion of ICTs promises to bring about greater global communication, and thereby greater trade – perhaps even greater mutual understanding, perhaps even more democracy around the world.
- On the other hand, the global diffusion of ICTs threatens to enrich those who are already well off, and to impose – often without awareness and thereby without choice – the cultural values and communicative preferences of one set of cultures upon others.

As a strict consequentialist, what specific benefits of this global diffusion can you point to (and, even better, support with documented evidence)? By contrast, what specific costs can you point to (also, ideally, supported with documented evidence)?

B. How do you assign relative *weights* to these benefits and costs? For example, you might have argued in (1) something like the following:

(i)   that the aggregate benefit of greater economic prosperity to the world population at large is of greater value than preserving some aspects of some "other" cultures, or

(ii) that whatever pleasures may follow for some (almost certainly, not all) in the global metropolis, as trade and wealth expand – such pleasures are not outweighed by the negative costs of threats to, and perhaps partial loss of, personal and cultural identity.

Either way, how do you (we) *decide* which set of possible consequences outweighs the other?

3. [*Deontology.*] As noted above, the United Nations' Declaration of Human Rights makes clear that individuals and cultures have a *right* to identity – i.e., to preserving, in our terms, the cultural values and communicative preferences that help define these identities. From this perspective, the global diffusion of ICTs, as fostering a covert form of cultural imperialism, thereby violates basic *rights*. We can add here that insofar as this imperialism is covert, it thereby further violates the basic (Kantian) insistence that only those acts that we *freely* choose are legitimate – in other words, such imperialism violates a basic *freedom of choice*.

(i)   What other sorts of *deontological* considerations come into play here – e.g., is there an equal and absolute right to free trade on the part of the wealthy nations that might be seen to conflict with an absolute right to individual and cultural identity?

(ii) What sorts of *deontological* resolutions might you be able to apply here?

Consider, for example, the *procedural* approach developed especially by Jürgen Habermas and, as applied specifically to information ethics, by Bernd Carsten Stahl (2004). On this approach, briefly, the norms and rules that we follow are legitimate as we are able to freely *choose* these – as these are debated and discussed in an *ideal speech situation*, in which

(a) every person affected by such norms and values is genuinely free to speak and argue;
(b) every participant practices *perspective-taking*, i.e., the effort to understand not only the logical arguments but also the emotional dimensions of other persons' claims; and
(c) agreement results from "the unforced force of the better argument" (Gimmler 1996).

So: imagine a conversation/debate over the values and rules that should guide the global diffusion of ICTs, with a focus on the possible dangers to individual and cultural identity as this diffusion might impose specific cultural values and communicative preferences – perhaps unawares – on those who are already at a disadvantage in the global marketplace.

(a) What sorts of values and/or rules for such diffusion might a person from a developed/culturally dominant country initially insist upon/agree to?
(b) What sorts of values and/or rules for such diffusion might a person from a developing/culturally marginalized country initially insist upon/agree to?
(c) What other voices/perspectives should be brought to this conversation/debate (e.g., representatives of future generations? of eco-systems? etc.)

Given these diverse voices and perspectives, can you discern and articulate a set of values and rules for the global diffusion of ICTs that all participants might agree to on rational and emotional grounds?

4. [*Virtue ethics.*] From the standpoint of virtue ethics, some of our primary ethical questions are: "how do I become a good human

being?" and "how do I conceive of and pursue the good life?" "The good life" here means primarily a life that, in concert and collaboration with others in the human and (for some – e.g., some feminists, Buddhists, Confucians, many indigenous peoples, etc.) larger biological sphere, and, indeed, with the larger cosmos itself, allows me to cultivate my best human capacities – which should result in a *harmonious* way of life and thus happiness.

Also remember: "virtue" in *virtue ethics* refers to the ancient Greek sense of *arête* or *excellence*. Virtue ethics is thus the ethics of becoming an *excellent* human being – and the *virtues* or excellences that I ought to pursue are thus the ones that help me achieve such excellence.

(i) What might a "virtue ethics" of ICT use – and diffusion – "look like"? To help arrive at this, try the following:

(a) List the sorts of *virtues* – practices, especially as these become habitual for us –that might contribute to the good life.

[*Hint*: "virtue lists" are easy enough to come by – e.g., Benjamin Franklin's list in his *Autobiography* (an online version is available at <http://www. earlyamerica.com/ lives/ franklin/chapt8/>). Perhaps with the help of your instructor, find and review a representative sample of these lists, including lists from Buddhist, Confucian, African, and/or other traditions. Then consider which among these virtues you think/feel belong to those that help human beings become *excellent*, both as individuals and as members of a larger community/communities – including a *global* community.]

(b) Develop a "log" of how you use ICTs and other digital media – e.g., everything from communication in various forms (SMS-/text-messaging and phone calls, email, surfing the web, spending time on a social network site, downloading music and videos, etc., etc.) as well as other activities made possible by digital media (photography, video games, etc.)

(c) Given your initial list of virtues, and then the various activities facilitated by ICTs and digital media, consider which of these virtues/excellences might be fostered – and/or countered – by various forms of ICT use.

(ii) Given this analysis, how would a virtue ethics approach evaluate and respond to the possible dangers of a covert imperialism brought about by the global diffusion of ICTS (and other digital media)?

That is:

(a) Might there be specific virtues/excellences that such diffusion could foster and enhance (e.g., our efforts to better understand one another, to develop more harmonious relationships with one another and our larger world, etc.)?

(b) Might there be specific virtues/excellences that such diffusion could counter and repress?

5. (*Additional ethical frameworks.*) After you have become familiar with one or more of the additional ethical frameworks discussed in chapter 6 – and/or as introduced by your instructor – apply one or more of these to the possible problems of the global diffusion of ICTs.

6. (*Meta-ethical discussion*). Especially as you have developed relatively complete responses to the possible problems of the global diffusion of ICTs (i.e., questions 2–5), now what?

That is:

A. Which of these theories results in responses to these problems that seem *closest* to your own initial intuitions and responses to these problems?

B. Which of these theories results in responses that seem to most sharply differ from, and/or contradict, your own intuitions and responses?

C. How might you resolve the difference/contradictions here?

Additional research/writing

1. The digital divide

Much has been written and researched regarding the "digital divide" – i.e., the disparities between what are sometimes called "the information rich" and "the information poor." Broadly, contrary to initial hopes that the global diffusion of ICTs would lead to greater global prosperity – the "rising tide that raises all boats" – much of the available data and analyses seem to show the contrary: the disparity between rich and poor is growing, not shrinking, both *within* such countries as the United States, and between the rich and poor nations of the world.

Research: what kinds of resources can you find to help illuminate the question, "Has the global diffusion of ICTs led to a broader global prosperity, one that has genuinely benefited both rich and poor – or has this diffusion served as just one more example of the rich getting richer while the poor get poorer?"

[*Hint*: a useful start is the United Nations Development Programme's "Statistics of the Human Development Report," <http://hdr.undp.org/en/statistics/>. In addition to a wide range of statistics relating to "the human development index" (with countries categorized in terms of high, middle, and low HDI), a search feature allows you to look up articles specifically referring to the digital divide.

The United Nations Development Programme's ICT for Development (ICTD) Observatory collects current news reports, organized by topics such as capacity building, the digital divide, etc.: <http://www.sdnp.undp.org/observatory>.]

Additional readings

Within the frameworks of intercultural information ethics:

Canellopoulou-Bottis, Maria and Kenneth Einar Himma. 2008. The Digital Divide: A Perspective for the Future. In Kenneth Einar Himma and Herman T. Tavani (eds.), *The Handbook of Information and Computer Ethics*, 621–37. Hoboken, N.J.: John Wiley and Sons. [Provides an excellent overview of the current, sometimes overwhelming, realities and understandings of what the "digital divide" means, and explores a number of ethical responses.]

Himma, Kenneth Einar. 2007. The Information Gap, the Digital Divide, and the Obligations of Affluent Nations. *IRIE International Review of Information Ethics* 7 (09). <http://www.i-r-i-e.net/inhalt/007/07-himma.pdf>, accessed September 19, 2008.

[Reviews the correlations between poverty, information gaps, and the digital divide, and argues for positive obligations on the part of affluent nations to help those in poverty.]

Hongladarom, Soraj. 2007b. Information Divide, Information Flow and Global Justice. *IRIE International Review of Information Ethics* 7 (09). <http://www.i-r-i-e.net/inhalt/007/08-hongladarom.pdf>, accessed September 19, 2008.

[Takes up a notion of global justice to address the problems of the digital divide – arguing, however, for an "intra-Southern" information flow to supplement a needed North–South information flow for the sake of development.]

Mesbahi, Mohamed. 2007. The Third World and the Paradox of the Digital Revolution. *IRIE International Review of Information Ethics* 7 (09). <http://www.i-r-i-e.net/inhalt/007/04-mesbahi.pdf>, accessed September 19, 2008.

[Provides an informed and thereby skeptical perspective on how far the promises of "the digital revolution" have been fulfilled in the third world.]

Finally, several readings from the recent anthology edited by Rafael Capurro, Johannes Frühbauer, and Thomas Hausmanninger (eds.), *Localizing the Internet: Ethical Aspects in Intercultural Perspective* (Munich: Fink Verlag, 2007), address various aspects of the digital divide, e.g.:

Jackson, Willy and Issiaka Mandé. 2007. "New Technologies" and "Ancient Africa": The Impact of New Information and Communication Technologies in Sub-Saharian [*sic*] Africa. 171–6.

Britz, Johannes. 2007. The Internet: The Missing Link between the Information Rich and the Information Poor? 265–77.

Finquelievich, Susana. 2007. A Toolkit to Empower Communities in Latin America. 301–19.

German readers will find the earlier anthology also very useful indeed:

Scheule, Rupert M., Rafael Capurro, and Thomas Hausmanninger (eds.). 2004. *Vernetzt gespalten: Der Digital Divide in ethischer Perspektive* [Networked/Split: Ethical Perspectives on the Digital Divide]. (Schriftenreihe des International Center for Information Ethics, Bd. 3). Munich: Wilhelm Fink. (For a helpful review, see Richard A. Spinello,

"Review: Vernetzt gespalten: Der Digital Divide in ethischer Perspektive," *International Review of Information Ethics*, 4 (2005), <http://www.i-r-i-e.net/inhalt/004/Spinello.pdf>, accessed September 19, 2008.)

2. Cross-cultural communication – offline/online

If you and/or your institution have access to Wolfgang Donsbach (ed.), *International Encyclopedia of Communication* (available online at <http://www.communicationencyclopedia.com/public/>, accessed September 19, 2008), the following articles, as their titles suggest, focus on important elements of communication within specific cultural domains:

Alemán, Carolos Galvan. 2008. Communication Modes, Hispanic.
Baldwin, John R. 2008. Communication Modes, Western.
Hecht, Michael L. 2008. Communication Modes, African.
Moghaddam, Fathali M. 2008. Communication Modes, Muslim.
Zhang, Yan Bing. 2008. Communication Modes, Asian.

Either individually and/or in groups, read and review at least two of these articles with care. In doing so:

A. Develop a list of important *commonalties and differences* in the communication styles characteristic of (at least two) different cultural domains.

B. Develop a list of suggestions for what someone coming from one of these cultural domains should do in his/her communication with someone from the "other" cultural domain, in order to avoid potential misunderstandings, insults, etc.

C. Given especially the important *differences* between these communicative styles, what are the *ethical* consequences?

(i) That is, is there an ethical obligation to learn how to communicate with one another in ways that respect and foster these differences?

If so, on what grounds? That is, what *arguments/ evidence/reasons/intuitions*, etc., can you offer in support of such an obligation?

If not, why not? Again, what *arguments/evidence/reasons/ intuitions*, etc., can you offer against such an obligation?

(ii) In light of these differences, are there specific *virtues* that would help communicants from one cultural domain better/more respectfully communicate with those from the other cultural domain?

If so, what are these?

[*Hint*: I have highlighted in the discussion above the virtues of humility, compassion, understanding, and forgiveness. You may want to consider each of these individually to determine whether or not they would in fact be useful in the specific case of cross-cultural communication that you are examining here.

This also means: you may discern other virtues that would also be useful.]

D. Given the possible dangers of the global diffusion of ICTs – dangers that follow in part because most designers and users of these technologies are not usually *aware* of the potential conflicts between diverse cultural values and communicative preferences (i.e., between those embedded and fostered by a specific set of technologies, and those prevailing and defining a "target" culture) – are the persons who take up and use these technologies in any way *responsible* for being aware of these possible dangers?

(i) What would that responsibility "look like"? That is, what specific sorts of things would a responsible user of ICTs attempt to be aware of, do, etc.?

(ii) How would you *justify* the claims that:

(a) yes, users of ICTs are responsible for being aware of the possible consequences of their uses of these technologies – including their possible impact on "others"?

(b) no, users of ICTs have no special responsibilities for cultivating such awareness?

For further reading

On cross-cultural communication: Hall, Hofstede, and "the Other" online

For classic readings:

Hall, Edward T. 1976. *Beyond Culture*. New York: Anchor Books.
Hofstede, Gert. 1980. *Culture's Consequences: International Differences in Work-Related Values*. Beverly Hills, CA: Sage.
—— 1983. National Cultures in Four Dimensions. *International Studies of Management and Organization* 13, 52–60.
—— 1984. The Cultural Relativity of the Quality of Life Concept. *Academy of Management Review* 9, 389–98.
—— 1991. *Cultures and Organizations: Software of the Mind*. London: McGraw-Hill.

For cultural analyses of online communication via the frameworks developed by Edward T. Hall and Gert Hofstede – including discussion of the limitations of these and emerging new frameworks, see:

Ess, Charles and Fay Sudweeks. 2005. Culture and Computer-Mediated Communication: Toward New Understandings. *Journal of Computer-Mediated Communication* 11 (1), article 9. <http://jcmc.indiana.edu/vol11/issue1/ess.html>, accessed September 19, 2008.

For analyses of cultural dimensions of ICTs with a specific focus on *indigenous peoples*, see:

Dyson, Laurel, Max Hendriks, and Stephen Grant (eds.). 2007. *Information Technology and Indigenous People*. Hershey, PA: Information Science Publishing.
Ess, Charles and Fay Sudweeks. 2001. On the Edge: Cultural Barriers and Catalysts to IT Diffusion among Remote and Marginalized Communities. *New Media and Society 3* (3), 259–69.

For a discussion of the ethical dimensions of establishing trust and respect online, see:

Weckert, John. 2007. Giving and Taking Offence in a Global Context. *International Journal of Technology and Human Interaction 3* (3), 25–35.

On the "Otherness of the Other":

Levinas, Emmanuel. 1963. La Trace de l'autre [The Trace of the Other]. A. Lingis, trans. *Tijdschrift voor Philosophie* (Sept.), 605–23.
[Much of the discussion of "Otherness" or alterity begins here.]
——. 1987. *Time and the Other and Additional Essays*. Richard A. Cohen, trans. Pittsburgh, PA: Duquesne University Press.

In regard to information ethics:

Hausmanninger, Thomas. 2007. Allowing for Difference: Some Preliminary Remarks Concerning Intercultural Information Ethics. In Rafael Capurro, Johannes Frühbauer, and Thomas Hausmanninger (eds.), *Localizing the Internet: Ethical Aspects in Intercultural Perspective* (Schriftenreihe des ICIE, Bd. 4), 39–56. Munich: Wilhelm Fink Verlag.
[A highly nuanced but clearly written philosophical analysis of diverse understandings of difference and otherness vis-à-vis information ethics and cultural pluralism.]

For discussion of the risks of computer-mediated colonization and the correlative ethical responsibilities of ICT designers (including, e.g., web-page designers):

Ess, Charles. 2006a. Du colonialisme informatique à un usage culturellement informé des TIC. In Joëlle Aden (ed.), *De Babel à la mondialisation: apport des sciences sociales à la didactique des langues*, 47–61. Dijon: CNDP – CRDP de Bourgogne.
—— 2006b. From Computer-Mediated Colonization to Culturally-Aware ICT Usage and Design. In Panayiotis Zaphiris and Sri Kurniawan (eds.), *Advances in Universal Web Design and Evaluation: Research, Trends and Opportunities*, 178–97. Hershey, PA: Idea Publishing.
—— 2006d. Universal Information Ethics? Ethical Pluralism and Social Justice. In Emma Rooksby and John Weckert (eds.), *Information Technology and Social Justice*, 69–92. Hershey, PA: Idea Publishing.
——. 2007c. Déclinaisons culturelles en ligne: observation « de l'autre ». Special issue of *Études de Linguistique Appliquée*. Paris: Didier Klienkensick. *D'autres espaces pour les cultures*, ed. Clara Farrao. No. 146 (avril–juin), 149–60.

On the Muhammad Cartoons

Debatin, Bernhard (ed.). 2007. *The Cartoon Debate and the Freedom of the Press: Conflicting Norms and Values in the Global Media Culture/Der Karikaturenstreit und die Pressefreiheit: Wert- und Normenkonflikte in der globalen Medienkultur.* Berlin: LIT Verlag 2007.
[In his book, Bernard Debatin includes a very helpful "Cartoon Controversy Timeline" (2006, pp. 227–32) that provides considerable detail of the various responses to the publication of the cartoons.]

For Scandinavian perspectives (my thanks to Prof. Knut Lundby, Department of Media and Communication, University of Oslo, for these suggestions):

Christoffersen, Lisbet (ed.). 2006. *Gudebilleder – Ytringsfrihed og religion i en globaliseret verden* [*Images of God: Freedom of Speech and Religion in a Globalized World*]. Copenhagen: Tiderne Skifter.
"*Dialog eller konfrontasjon?* [Dialogue or Confrontation?]", <http:// www.culcom.uio.no/aktivitet/karikaturene.html>, a conference and project of "CULCOM: Kulterell kompleksitet I det nye Norge [Cultural Complexity in the New Norway]".
Kunelius, Risto, Elisabeth Eide, Oliver Hahn, and Roland Schroeder. 2007. *Reading the Mohammed Cartoons Controversy: An International Analysis of Press Discourses on Free Speech and Political Spin.* Bochum, Germany: Projekt Verlag.

CHAPTER 5

Still More Ethical Issues

Digital Sex and Games

Sex shaped the Internet as it exists today.

(Perdue 2004: 260)

Chapter overview

Following a "snapshot" of how pornography and violence appear to predominate in at least popular media coverage of digital media – meaning, the Internet, cellphones, and computer games – we consider how these more popular approaches tend to include "moral panics" that may hinder our ethical analysis. In particular, we will explore the logical matter of *exclusive* vs. *inclusive* "ors." We will then examine some of the more central ethical issues in terms of *utilitarianism, deontology, feminist ethics/ethics of care,* and *virtue ethics.* (If you have not already reviewed these frameworks for ethical decision-making in ch. 6, you should do so before moving into this chapter.) The chapter includes a specific analysis of online pornography vis-à-vis feminist and post-feminist perspectives, followed by a similar analysis of violence and videogames from a virtue ethics perspective.

Introduction: sex, games – and "the media" on digital media

It seems that daily there are yet more reports like the following. First:

> A local pastor is captured in an online sting operation. An "Internet detective" – usually a law enforcement officer, but occasionally Internet vigilantes (e.g. "Catch a perv" – Silver 2006) – posing as a young girl in a chat room entices an older

man, who hopes to meet up with the "girl" for sex, into revealing his identity. Minimally, profound embarrassment follows – often along with legal charges and penalties (Saavedra 2008).

Second, MySpace earnestly promises to do more to protect young children from sexual predators and pornography:

> MySpace, which says it has about 70 million users, agreed to install safeguards that require an adult user to prove that he or she knows a child user in order to contact that child, for instance by typing in an address or phone number.
>
> Profiles of users under 18 will automatically be set to "private," preventing casual browsers from seeing them.
>
> Parents who do not want their children using the site will be able to submit their children's e-mail addresses to MySpace, which will prevent users of those addresses from creating profiles. MySpace will hire a contractor to identify and remove pornographic images and links to pornographic sites from its Web site.

But such measures, of course, can be easily gotten around:

> Children can create new e-mail addresses that their parents do not know about; adult strangers could obtain enough information about children to get past the site's safeguards; pornographic links spring up as quickly as they can be removed. (Barnard 2008)

Finally, what will happen now that airline passengers (in the U.S.) have Internet access?

> Seat 17D is yapping endlessly on an Internet phone call. Seat 16F is flaming Seat 16D with expletive-laden chats. Seat 16E is too busy surfing porn sites to care. Seat 17C just wants to sleep.
>
> Welcome to the promise of the Internet at 33,000 feet – and the questions of etiquette, openness and free speech that airlines and service providers will have to grapple with as they bring Internet access to the skies. (Jesdanun 2007)

Apparently, all hell will break loose . . .

Sex and games in digital media

Beyond these thematic examples of daily media coverage, more broadly, pornography online remains a staple of the Internet (and other forms of digital media), contributing, by some estimates, between 30 and 40% of the total revenue streams generated by online commerce.

But of course, there's not only every imaginable form of sex and sexuality available online; there's more sex – and violence – in videogames. Game proponents point out, for example, that in Germany, only ca. 3% of all games can be classified as "adult only" – with the rest made up of a range of games, including educational games for children (*Networld* 2005). Denmark, in particular, enjoys a strong international reputation for conjoining electronic games with education – including education about basic democratic values (Tasha Buch, personal communication, November 23, 2007). More broadly, electronic games may offer distinctive advantages in education (e.g., Gee 2003). And, of course, *play* is unquestionably important in human life – so much so, for example, that Johan Huizinga famously named us *Homo Ludens* – "[hu]man the player" ([1938] 1955).

But, whether fairly or not (a topic we'll return to), the Columbine (Colorado) killings in 1999 were linked with (among other things) the killers' affection for violent videogames. Subsequent shootings – not only in the United States (e.g., the Virginia Tech killings in 2007) but also, e.g., in Emsdetten, Germany (2006) and Jokela High School in Tuusula, Finland (2007) – are likewise connected in media reports with a focus on the killers' favorite violent computer games.

Nor is the concern with sex, violence, and digital media limited to videogames: mobile phones equipped with video recording and playback are also a focus of concern. As we saw in chapter 2, some Danish youths used their phones to video-record their sexual encounters, apparently without their partners noticing. More darkly, among some young German males, cellphones were used for storing and playing "snuff-videos" that depict, e.g., a Russian soldier's throat being sliced. In addition, cellphones are central to the process (game?) of "happy slapping" – of staging a fight between two willing cohorts, while others record the violence on their cellphones for later replay and distribution. The staging, moreover, may go beyond friends: sometimes, a passer-by is selected as a victim for assault – again, all video-recorded for later entertainment (Kahlweit 2007; Picturephoning.com n.d.).

All of this is embroiled with national efforts to protect children –
and, in some ways, adults – from various harms (real, potential, and
disputed) that may well come from these various uses of digital
media. These include various rating systems designed to determine
(at least roughly) what level of sex and/or violence (if any) is appropriate for a person of a particular age to consume. In the U.S. case,
for example, the Entertainment Software Rating Board (ESRB) has
settled on seven rating symbols, resulting in more than 25 possible
labels that warn consumers regarding the content of a given
videogame – certainly including sex and violence, but also adult
language, drug use, gambling, and other worrisome behaviors
(<http://www.esrb.org/ratings/ratings_guide.jsp>). Germany has
even stricter laws forbidding the sale of "killer-games" (*Killer-Spiele*)
to young children. Indeed, German efforts have included recruiting
children in the fight against illegal sales of violent videogames to
children – by asking them to make "test purchases" (*Testkäufe*) of
such products in various stores. Critics decry the effort, noting that
"children are not bait [*Kinder sind keine Köder*]" (*Die Welt* 2007: 1).

Finally, to state the obvious: countries have attempted to limit
access online to pornography in various ways. Broadly, the debates
swirling around what *is* pornographic, what harms (if any) it entails,
and thus if and, and if so, how it is to be regulated likewise vary
widely from country to country. Denmark, for example, is famous
for first legalizing what had been forbidden forms of pornography
in 1969 (*Time* 1969). In recent years, so-called "hardcore" pornography[1] has become freely available on a Copenhagen cable channel –
one phenomenon contributing to what some have called the
"pornification" of society (e.g., Paul 2005: Paasonen et al. 2007a).
Suffice it here to say that in the U.S. and China, by contrast, prevailing attitudes – and correlative laws – are considerably more
restrictive. But the Internet, of course, makes it possible to communicate across national boundaries – what is illegal in one country

---

1 Definitions are notoriously tricky here – but in this case, "hardcore" includes full
depiction of various sex acts, including group sex, sado-masochism, etc., in contrast with "softcore" pornography, which tends to remain content with partial
nudity and suggestive poses.

may be perfectly legal and accessible from another country. And how far various regimes will be able to detect and control what they deem to be inappropriate information (whether pornography, news and opinion unflattering to a particular country/regime, etc.) remains an on-going matter of discussion and development. U.S. sites such as eBay, for example, were blocked from German and French users until eBay restricted the sale of material forbidden in those countries – primarily, Nazi paraphernalia. In the meantime, roughly analogous to the U.S. Entertainment Software Rating Board (ESRB), the Internet Content Rating Association (ICRA) – initially working in both the U.S. and the United Kingdom (e.g., with the British Board of Film Classification), and now part of the "Family Online Safety Institute" – hopes to protect children from the potential harms of violence and pornography online through a voluntary system of industry self-regulation: " . . . self-regulation leads to the best balance between the free flow of digital content and protecting children from potentially harmful material" (<http://www.fosi.org/icra>). But while the ICRA claims to cross cultures, as we have seen (ch. 2, privacy), our views as to whether such matters are better settled through national legislation or industry self-regulation are themselves in part culturally variable: the former prevails in the European Union, while the latter prevails in the U.S., for example.

The only ethical "bright line" that (most) individuals and countries have agreed upon in these arenas is with regard to child pornography. But, of course, as in all things ethical – and certainly all things sexual – even this, along with all the other matters, is contested in an array of on-going debates.

Before we turn to ethics, two caveats . . .

These topics are hence profoundly complex – most especially as we attempt to assess and respond to them in ethically informed ways. Broadly speaking, however, it helps to sort out the arguments in terms of the primary ethical decision-making frameworks that we bring to bear on such matters, beginning with *utilitarianism*. Before doing so, however, a final observation and two caveats. Consider the subheading for one of the articles cited above:

Two students have set up an internet chatroom that catches out men making sexually explicit advances to someone they believe is a 13-year-old girl. Entrapment, or a public service? ("Catch a perv" – Silver 2006)

First problem: exclusive "ors" and questionable dichotomies

Notice the last "or" here – an *exclusive* "or," logicians would call it: either this activity counts as an unfair act against the victims whose identities are revealed through the students' efforts, or these revelations are a vital and important public service (but not, presumably, both). This either/or approach, you should notice, dominates much of "the media" reporting surrounding a range of topics, including new technologies (e.g., genetically modified foods), especially digital media, and most especially matters of sex and violence vis-à-vis the Internet and videogames (cf. Buchanan and Ess 2005).

There are two particular problems here that are important for us to straighten out from the outset. One is the *logical* matter. However tempting and common it is for us to think in terms of *exclusive* either/ors (one thing can be true – or the other thing can be true – but both can't be true simultaneously), in point of fact there is a second kind of "either/or": the *inclusive* either/or, so-called because it includes (rather than excludes) the possibility of two things being true simultaneously.

Again, it may be tempting to think that an act – such as the intentional baiting and deception of the two students described above – is *either* a *good* thing (public service) *or* a *bad* thing (entrapment), *but not both*. And thinking in this way may not only be common for journalistic headlines, but also may reflect a particular stage in our cognitive development (i.e., some evidence suggests that thinking dualistically is more characteristic of younger persons, with movement towards greater use of "both/and" ways of thinking in late adolescence and early adulthood). But this emphasis on an exclusive either/or is also characteristic of the meta-ethical position of ethical absolutism, as described in chapter 6: *either* a given value or act is right, or it is wrong, but not both. And: either people agree in their beliefs and actions with a single set of what one holds to be universal moral values, or they are wrong, but not both.

To be sure, there are realities that are perfectly well described in terms of the exclusive "or": a basic light, for example, is either on, or off, but not both. And *within* the ethical perspective of the ethical absolutist, there are certainly some norms and acts that are right – and so *other* (different) norms and acts are *wrong*. As an easy example: social and religious conservatives, if they are absolutist in this way, would hold heterosexuality, and that, solely within the bounds of marriage, as the *right* set of sexual norms and practices; anything different from those hetersexual norms and practices – to use the acronym, anything GLBT (Gay, Lesbian, Bisexual, Transgendered) – is hence wrong.

This does not mean, of course, that only social and religious conservatives are ethical absolutists, and that only they may think in terms of exclusive "ors." On the contrary, I find that all of us are ethical absolutists about one or more important set of norms and values – and all of us will thereby invoke an exclusive "or" on one or more topics. As another simple example: social and religious progressives in the United States and Europe insist that, at least up to a point, a woman has a *right* to choose with regard to continuing or aborting a pregnancy. Insofar as this belief is held in absolutist fashion, then those who disagree (obviously, religious and social conservatives) are wrong.

But as I hope the example "entrapment or public service" suggests, especially in ethical matters, we are frequently confronted *not* by simple, exclusive either/ors, but rather by inclusive ones. In this case: why can't the same act be *both* entrapment *and* a public service?

This is at least a question we should always ask when confronted with what appears to be an exclusive either/or: is it possible, in fact, that both alternatives can be true simultaneously?

Hence, we need to be wary. While journalistic practice (if not our own absolutism) may favor simple and exclusive either/ors, we should be careful not to let that practice overly shape our own thinking. Before accepting/assuming that the only ethical choices before us are indeed those of an exclusive either/or – especially if we are influenced in our thinking by journalistic practices that frame the issues for us in this way – we should make every effort to see if *both* possibilities may be true. Indeed, we should go on to see if there might still be other legitimate alternatives not even expressed in these dichotomies – again, to make sure that our thinking is not unnecessarily (and unfairly) limited. (This ability to think beyond the choices initially presented to us is not only a mark of critical and independent thinking – one that helps us see the possibilities of more than one claim being true; it may also help us break through what logicians rightly call "false dichotomies." Such dichotomies are false precisely because they exclude options that should be considered.)

Problem two: moral panics as framing both critique and defense

The second problem is that this way of couching important matters – while certainly good for creating excitement and selling newspapers (and other media) – often goes hand in hand with creating "moral panics." Moral panics have been studied closely since the 1970s, but with examples including the introduction of the then-new technologies of the automobile and radio in the 1920s. For our purposes, the main point about moral panics is that they entail reactions of moral outrage in response, for example, to a new technology (as engendering moral decay), new immigrant groups (whose moral values and practices are clearly wrong in the eyes of the dominant society) – and frequently, it appears, the behavior of young people(!). They are fueled by media stories that strongly sensationalize – in part, as they strongly *polarize*, precisely by casting the emerging phenomena in dualistic terms of good (prevailing norms) and bad (emerging behaviors as threatening those norms: see Critcher 2006).

To be sure, sometimes strong ethical and moral reactions in the face of some novelties might be justified. And, just as I have done to open this chapter, it is sometimes effective to highlight striking, if extreme, cases, simply to get our attention in an initial way.

But moral panics can mislead us, not only as they frame important issues in too simple dichotomies; they also make it too easy to *reject* what may be important ethical critiques as we seek to ignore those critiques as *only* the result of a moral panic. For example, if we ourselves like to play videogames, we might want to defend them against the criticisms highlighted in media accounts that suggest, for example, that violent videogames are dangerous to society because they inspire real-world violence. And so we might then be sorely tempted to dismiss those claims as simply one more instance of a moral panic that we can then safely ignore. But this is just that – too simple. In order to reject claims such as these, as highlighted in media accounts, then the burden of proof is on us to show that these emerging norms and behaviors are somehow justified. (Don't worry – you'll have your chance to do so!)

[At this point, all I want to convey is that as we try to take up these important issues in a careful and systematic way, we should not let our thinking about them be framed from the outset solely in the exclusive "either/or" terms that a good deal of popular media use in presenting these new technologies and the (sometimes) new behaviors that they make possible, especially as these may be feeding into a moral panic.]

Ethical analyses – a first look

In terms that should now be familiar to us, we can understand the ethical dimensions here initially in *utilitarian* terms. On the one hand, new technologies and their various applications are attractive to us because of clear and/or promised *benefits*. As an initial but by no means comprehensive list, some of the benefits used as ethical reasons in favor of using these technologies and applications include:

- *cellphones, ICTs* – greater convenience, speed and reach in communicating with others, both locally and around the world; new and valuable social networking possibilities, etc.
- *games* – educational benefits from serious/educational games, including development of new skills taught via simulation and role-play; entertainment, relaxation, and play as essential elements in life, etc.
- *pornography* – may serve as a venue of free expression and exploration of sexual identity and/or form of sex education; harmless

stimulation and/or sexual release that might express itself otherwise in negative and harmful ways; etc.

And for all of these, certainly, proponents will point to *economic* benefits as these new technologies and applications open up new possibilities for businesses, large and small.

[*Nota bene*: if you don't see here *your* particular supporting reason(s), it may be because (a) this is not intended to be a complete list, and/or (b) your supporting reasons stem from a different ethical framework. We'll get there . . .]

On the other hand, what obviously vexes parents, teachers, cohorts, and states are a range of *harms* – both to individuals and to the larger society – that may also follow. Again, an initial but, as we shall see, by no means comprehensive list:

- *cellphones, ICTs* – dangerous or at least useless distractions; nuisance to others; "too much information" – too easy access to forms of information that one should not/does not need (including, e.g., pornography and child pornography; stalking and sexual predation); addictive; vast potential for violating, eliminating privacy; will lead to moral and social decay, etc.
- *games* – may be addictive; encourage violent behavior; will lead to moral and social decay, etc.
- *pornography* – may be addictive; stimulates aggressive attitudes/actions towards women as objects; may develop attitudes/behaviors among men that make it more difficult for them to develop and sustain relationships with women; will lead to moral and social decay, etc.

[*Nota bene*: again, if you don't see here *your* particular objection(s), it may be because (a) this is not intended to be a complete list, and/or (b) your objections stem from a different ethical framework. We'll get there . . .]

[In this first look, it is important to notice (as may be obvious) that several of these (real/potential) harms presume a *causal* connection between exposure and immersion in sexual and/or violent images and real-world behavior. Given such a connection, states would then indeed have an obligation to protect their citizens – especially their younger and more vulnerable citizens – from harm. Whether or not such harms can be demonstrated, as we will see, is highly contested.]

Initial reflection/discussion/writing exercise: media, ethics, and "the media"

1. Describe your own consumption/use of digital media – including using MP3 players, mobile phones (talking, text-messaging, photo-recording and playback – including watching TV, etc.), the Internet (email, research, surfing websites, shopping, etc.), and videogames.

2. How do your patterns of consumption/use compare with your cohorts? That is, are you a relatively light/moderate/heavy user of these technologies?

3. Develop an initial list of what you see as the ethical pros and cons of your usage of these media. (You might review the lists given above, just as a starting point.)

    A. In your view, what are the most important ethical reasons *for* your using these media in the ways you've described?
    B. In your view, are there possible ethical reasons that would argue *against* your using these media in the ways you've described? If so, which do you see as the most important?

4. Has someone ever criticized you for a specific use of these media – because of one or more *ethical* reasons?
    If so,

    (i) describe the specific use, and
    (ii) state as carefully (and fairly) as you can what the ethical objection was (or objections, if there was more than one).

5. (*Meta-theoretical analysis.*) As you review the list of ethical reasons, both pro and con (developed in (3) and (4)) for a given use of digital media, can you tell which of these derive from

(i) *utilitarian* approaches (e.g., benefits outweighing costs or harms – and/or vice versa);

(ii) *deontological* approaches (e.g., emphasizing fundamental duties towards others, beginning with *respect* as free human beings; protection of basic rights, etc.);

(iii)*feminist ethics/ethics of care* (e.g., emphasizing equality of women and men, the important of grounding our ethical actions in care for others and with a view towards sustaining inclusive "webs of relationships");

(iv) *virtue ethics* (e.g., emphasizing developing the habits and practices that foster our excellence as human beings, including learning how to make sound ethical judgments);

(v) none of the above?

Pornography online: feminist, postfeminist, and global perspectives

The 1950s and especially the sexual revolution(s) of the 1960s began to call into question traditional sexual norms and practices, including restrictions on the production and consumption of pornography. In Europe, as noted above, Denmark was the first country to legalize pornography in 1969 (*Time* 1969); West Germany and Sweden soon followed suit (1970 and 1971, respectively: Paasonen et al. 2007b: 7). In the United States, the debates between social and religious conservatives and progressives circled around the polarities of preserving traditional morality and the importance of protecting women and children from harms associated with pornography vs. freedom of expression as protected by First Amendment (of the U.S. Constitution) rights. The First Amendment and its subsequent application in American life to defend especially artistic freedom of expression has led to a stronger emphasis on freedom of speech in the U.S. than in some other countries and traditions. Within this legal and cultural milieu, efforts to restrict access to pornography have long been attacked as censorship and thus as a violation of freedom of expression.

These debates were quickly complicated by further debate *within* the rising (second wave) feminist movement. On the one hand, feminists in general viewed traditional sexual norms and practices as grounded on gender stereotypes that denigrated women and justified their subordination to men. Hence, calling those practices and norms into question – including endorsing sex outside the bounds of traditional marriage – was one among many strategies intended to overcome the multiple forms of women's oppression. On the other hand, for the majority of U.S. feminists from the 1960s through the mid-1980s, pornography was emblematic of that oppression (e.g., MacKinnon 1989, 1992, 1993; see Adams 1996 and Cavalier 1996 for discussion). In this way, U.S. feminists distinguished themselves from religious conservatives, whose anti-pornography arguments derived from moral frameworks grounded in specifically religious worldviews, coupled with arguments regarding the putative harms of pornography on the consumer (Adams 1996).

But the "anti-porn feminists" were soon confronted by a further range of critiques – this time, from within their own camp. To begin with, the anti-porn feminists – represented most sharply by Andrea Dworkin and Catherine MacKinnon's efforts to develop definitions of pornography that would thereby strengthen its regulation and control (MacKinnon 1993) – thus found themselves siding with religious conservatives who also favored censoring pornography. By contrast, the "anti-anti-porn feminists" argued that such censorship could only be dangerous for women, especially progressive feminists, as well as for other groups marginalized by mainstream society (Cavalier 1996). More radically, anti-anti-porn feminists further called into question many of the basic assumptions underlying the anti-porn feminists' arguments. To begin with, the anti-porn feminists appeared to agree with – and thereby again reinforce – in a crucial way the assumptions of their conservative adversaries regarding the normativity of heterosexual sex. For the anti-anti-porn feminists, however, the original feminists' agenda of sexual liberation and equality must further mean overcoming the marginalization of "other" sexualities – including Gay and Lesbian, Bisexual, and Transgendered sexualities. In addition, especially as influenced by postmodernist conceptions of gender roles as "performance," any given set of sexual sensibilities and desires could be (perhaps only could be) seen as an *aesthetic* act, one that thus again falls under the category of freedom of expression. In short, from these perspectives, efforts on the part of anti-porn feminists such as Dworkin and MacKinnon – however well intended to protect women and children from ostensible harm and oppression – were seen in turn to be as repressive as their more conservative allies: censorship of "pornography" meant censorship of women's freedom to define and express their own sexualities above and beyond the boundaries and assumptions of heterosexuality as the norm (cf. Paasonen et al. 2007b: 17f.).

The U.S. context is important for us in part because it set the terms and frameworks for the early debates on pornography online.[2] And so, for example, an online dialogue on the issues of pornography, feminism, and censorship emphasized the three U.S.-centric positions of anti-porn feminists, anti-censorship feminists, and anti-anti-porn feminists (Cavalier and Ess 1996). But of course, as the Internet expanded rapidly beyond the borders of the United States, it thereby facilitated the representation of a still wider range of perspectives and voices in these debates. Hence the debates have moved beyond the earlier terms and approaches in important ways. To begin with, in a global society increasingly interconnected via the Internet and the World Wide Web, the definitions of what counts as "pornography" have only become more diverse (covering in India, for example, material that simply implies sex, including beauty pageants [Ghosh 1999]; in Indonesia, the term is bound up with laws regulating women's clothing and demeanor, including public displays of affection [Lim 2006, both cited in Paasonen et al. 2007b: 16]), thereby complicating the debates surrounding regulation of pornography online. And of course, responses and arguments further vary from country to country. For example, Paasonen and her co-editors contrast the U.S. and Nordic contexts, where the latter include a more permissive attitude

---

2 It may be helpful to remember here that as late as 1998, over 84% of those using the Internet were located in North America (GVU 1998). While regrettable from the standpoint of pluralism, the dominance of U.S. perspectives in these early debates is at least understandable in light of the Internet demographics at the time.

towards commercial sex, more "subdued conservative Christian perspectives," and an "equality feminism" that avoided at least some of the sex and culture wars that shaped debates in the U.S. (2007b: 15f.). As another reflection of more open attitudes outside the U.S., "netporn" – defined by Paasonen as "the ways in which online technologies restructure the pornographic, encompassing variations of peer-to-peer porn, amateur porn and [USENET group] alt.porn" (2007: 1) – has developed to the point of enjoying its own conferences in Amsterdam in 2005 and 2007. As described by Katrien Jacobs, Marije Janssen, and Matteo Pasquinelli, the netporn conferences

> . . . are novel zones for academics, activists and artists to discuss and expe-
> rience new phenomena around web-based sex and pornography. We are
> part of a porn-friendly, yet critical digital generation, bothered by a cultural
> climate of narrow-mindedness and porn hysteria. Critical studies about
> pornography and queer activism have been carried out in previous decades,
> but we are looking to discuss our tactile immersion in pornographic net-
> works. (2007: 1)

These and other emerging explorations of "pornography" should make clear that while the debates concerning pornography online were initially shaped by U.S.-based frameworks, those frameworks, as rooted in a complex U.S. history profoundly shaped by both Puritan beginnings and more recent Protestant funda-mentalism, vis-à-vis a vigorous tradition of free speech rights and distinctive feminist traditions, do not necessarily apply beyond U.S. borders.

The danger for us, as Paasonen and her co-editors point out, is that these North American approaches threaten to obscure and over-ride important alternatives and variations more openly expressed elsewhere. As a particular example: Pamela Paul's 2005 book, *Pornified: How Pornography is Transforming Our Lives, Our Relationships, and Our Families,* tries to avoid the polarities between what we might now think of as traditional feminist and religious conservative ("Puritanical") approaches that have dominated U.S. debate for the past several decades – while nonetheless arguing that in light of its harms, the consumption of pornography, like the con-sumption of cigarettes, can and must be regulated. But from

Paasonen and her co-editors' perspective, this approach remains comparatively limited: "Paul maps pornification in terms of compulsion and alienation while ultimately failing to account for the diverse aesthetics and practices involved" (2007b: 15). Moreover, the apparent alternative perspective in the U.S. only further contributes to a polarization of viewpoints characteristic of U.S.-based discussion: after considering alternatives to Paul in the U.S., Paasonen and her co-editors note: "Considered together, they make evident some dominant traits in public debates on pornography: while some discuss pornography as a social problem, others de-politicize it as hip and fun. There is little in terms of middle ground" (2007: 15).

Of course, some may well argue that there can *be* no middle ground on these (and other) issues. But at least at the outset, we want to be careful that we do not fall into potentially *false* dichotomies and too-easy exclusive "ors," especially as these may have dominated early U.S. debates and thereby threaten to dominate the views and debates outside the U.S. as well. More broadly, these important differences between diverse national and cultural perspectives on matters of sexuality and pornography mean that "[a]lthough certain global trends are recognizable, questions of regulation, policy and public opinion should not be generalized" (Paasonen et al. 2007b: 16).

Reflection/discussion/writing questions: ethical responses to porn?

Given this expanded view of possible perspectives on pornography:

1. Do you find one or more arguments, claims, insights, etc., as presented above that help strengthen your own position and views regarding pornography and its easy accessibility via digital media? If so, which one(s)?

2. Do you find one or more arguments, claims, insights, etc., as presented above that significantly *critique* or undermine your original arguments, claims, etc.? If so, which one(s)?

3. Are there any additional perspectives and arguments from your experience and reading that should come into play in ethically evaluating access to pornography (whether more "softcore" or more "hardcore") through contemporary digital media (certainly the Internet, but also through mobile phones)?

   If so, articulate these as clearly as you can – ideally, with references to one or more sources that you and/or your cohorts might use in an essay, for example.

4. By now, you should have a reasonably clear idea of what your preferred/familiar ethical frameworks are (*utilitarian, deontological, feminist ethics of care, virtue ethics,* and/or another perspective).

   And by now, you may also have identified one or more ethical issues involved with pornography as easily producible and accessible through digital media, such as possible harms to children, possible risks of addiction, possible harms to relationships, the dangers of censorship, the importance of free exploration and expression of one's own sexuality, etc.

   Identify one of these issues, and then show how your ethical framework(s) responds to these issues.

Case-study 1: virtual sex?

A number of futurists have predicted the realization of a more or less complete "telepresence" within the next decade or two. This would involve a diaphanous body-suit equipped with both micro-sensors and micro-stimulators that would allow for an experience of virtual sex online. That is, the suit would both send and receive information about the movement and response of the body as represented in an online virtual world as an avatar, and as one's avatar interacted with other such avatars likewise representing online the real bodies of others. (For additional readings, see Rheingold [1990] 2004; Barber 2004; Stuart 2008.)

Especially if one takes a *consequentialist* approach to ethics, such virtual sex could, on first blush (pun intended), dramatically change the ethics of sexuality. Specifically: if the major arguments *against* sex outside of marriage include warnings about the possible consequences – ranging from STDs to unwanted pregnancy – then most (but perhaps not all) such arguments evaporate with regard to virtual sex.

On the other hand, if one takes a more *deontological* approach – including attention, say, to promises of sexual fidelity made and thereby needing to be kept – such virtual sex would remain ethically problematic. In particular, as deontological ethics emphasizes the *intentions* of one's actions, the very thought of infidelity is ethically sanctioned. (So Jesus declared in his Sermon on the Mount: "'You have heard that it was said, 'You shall not commit adultery.' But I say to you that everyone who looks at a woman with lust has already committed adultery with her in his heart.'" Matthew 6.27–8, New Revised Standard Translation.)

Similarly, *virtue ethics*, *ethics of care*, and ethical frameworks that emphasize the importance of who we are in terms of the *relationships* we have and maintain with others might find strong reasons to condemn, say, virtual sex between strangers or multiple partners. By contrast, virtual sex between a married couple whose work and obligations keep them apart for long periods of time might be an important way of reinforcing psychological intimacy with at least a close version of physical intimacy – and hence ethically unproblematic.

Reflection questions

In light of these possibilities, and/or others not elaborated here but which may be important for your ethical reflections:

1. Articulate your beliefs and values regarding sexuality – specifically, when (if ever) sexual intercourse with another person is ethically justifiable.

   As you do so, provide the best *arguments*, *evidence*, and/or other supporting grounds or reasons for your views that you can.

2. Consider the possibility of having a virtual sexual encounter with another person via your avatars as shared through a future, very-high-bandwidth connection.

   (a) Offhand, under what circumstances (if any) might you consider such an encounter ethically justifiable?

   (b) Under what circumstances (if any) might you consider such an encounter as ethically unjustifiable?

   Again, what *arguments*, *evidence*, and/or other supporting grounds or reasons can you offer for your views?

Case-study 2: pornography and "pornification" of society

As we have seen, Denmark was the first Western country to legalize pornography (in 1969). This relatively relaxed attitude towards pornography seems to fit with a larger set of Danish values, including: tolerance, openness, and a characteristically northern European/Scandinavian lack of U.S.-style prudishness (if not religiously rooted guilt) about the human body and sexuality.

(In particular, in a major reform movement in Denmark in the nineteenth century, Lutheran theologian N.F.S. Grundtvig, in part influenced by more optimistic notions of human nature at root in the Enlightenment, eliminated the notion of *Original Sin* from his theology as part of a larger push towards gender equality, beginning with education for women through high school.[3] Other material developments likewise contributed to a growing gender equality, including expanding agricultural development in the nineteenth century, as this often required women to exercise greater responsibility and authority.)

Hence, both softcore and hardcore pornographic materials are easily available – including, as noted, the broadcast of hardcore pornography on a Copenhagen cable TV channel, one that is *not* scrambled in ways that would prevent casual (and/or young) viewers from stumbling across it.

Just as in the United States, a number of voices have protested against what they characterize as "the pornification of society," referring to the way in the forms and images of what were once considered taboo materials, produced explicitly for the purpose of sexual arousal, increasingly appear in our everyday lives – most especially in advertising, but also in mainstream entertainment, dress styles for young women, etc. These voices come largely from two, rather contrary directions. On the one hand, progressive and often (but by no means always) secular feminists see in such images (a) simply a continuation of old-fashioned and very destructive myths about

---

3   The doctrine of Original Sin is historically associated with patriarchal control of women: as the doctrine lays the responsibility for the introduction of sin and death into the world upon Eve, it thereby plays into the demonization of body, sexuality, and women in general. This interpretation of the 2nd Genesis creation story (Genesis 2.4–3.2), while orthodox in Western Roman Catholicism and subsequently among some Protestant reformers, is directly contrary to earlier Christian and Jewish readings of the text, which rather emphasize the positive nature of Eve's choice: acquiring "the knowledge of good and evil" is specifically understood as the acquisition of the distinctively human capacities of moral understanding and free choice – capacities that, in turn, early Enlightenment thinkers such as John Locke see as foundational to arguments for *democratic* polity, i.e., the political arrangements of human beings capable of rational self-rule.

women – myths that only contribute to their subordination – and/or (b) the continued objectification and commodification of women as simply sexual objects – practices that likewise seem to condemn women to the status of nothing more than meat (e.g. Adams [1990] 2000). On the other hand, religiously based social conservatives – whose views may, in part, be understood as resistance *against* the "anti-demonization" of women and sexuality by secular-progressive society, and thereby against the impulse towards gender equality – likewise decry pornography and pornification as one more sign of the moral decline of society.

In the Danish context, critics countered that both were wrong, arguing in part that sexually suggestive materials had been part of the public sphere since antiquity (see, e.g., the fresco from Pompei currently illustrating the Danish Wikipedia article on *"Pornoficering"*: <http://da.wikipedia.org/wiki/Pornoficering>). Given the extent to which such materials have been found in human cultures throughout history, the argument went, such materials are hardly the effect of the legalization and spread of pornography, but simply an expression of human *nature*.

Finally, Susanna Paasonen (email, February 29, 2008) points out that we should further distinguish between "pornification"

> . . . in the sense of soft-core aesthetics and references penetrating so-called mainstream popular culture (in music videos, advertising etc.), and the increased accessibility and perhaps even acceptability of hardcore porn.

She goes on to comment that while these two developments are closely connected, they may well have very different implications. She suggests, for example that

> . . . hardcore pornographies may involve some redefinition of the sexual as different fringes [i.e., beyond "normal" heterosexuality] become incorporated into the palette of commercial porn whereas the soft-core aesthetics à la Britney Spears do not necessarily involve any such thing.

Reflection questions

1. Does it seem to you that your society is "pornified"? That is, is pornography – in either hardcore and/or softcore form, including "soft-core aesthetics" of the sort identified by Paasonen – more and more pervasive? More and more accepted as normal forms of entertainment? If so – and depending which form(s) you identify (i.e., more hardcore or more

softcore) – does it seem to you that this pornification thereby plays some role in what Paasonen et al. call "the education of desire," i.e., in shaping our understandings, beliefs, expectations, and desires surrounding sexuality?

Be careful to keep in mind here Paasonen's distinction between hardcore pornography, as exposing us to what in the past would have been considered "fringe" sexualities, and softcore pornography, whose aesthetics show up, e.g., in music videos, various forms of advertising, etc. Again, Paasonen suggests that the former may redefine our understandings of sexuality in fundamental ways, while the latter may not. Hence, if "pornification" in your society is apparent more in the form of the emergence of "softcore" aesthetics in advertising, say, than in greater accessibility to and acceptability of hardcore pornography, then the potential impacts of such pornification will differ from those in which pornification includes more access and acceptability of hardcore materials.

Whatever your claim in response to this question, please provide one or more elements of evidence or support for your claim.

2. If there is a "pornification" of society taking place, how do you ethically analyze it? That is, do you see it, e.g., as ethically justified as it may make greater room for greater diversity and expression of sexuality, call into question traditional gender stereotypes that reinforce women's subordination to men, make available important forms of sex education, etc., and/or as leading to a destructive corrosion of important ethical and social values, reducing men's ability to form and sustain relationships with women, leading to greater violence and sexual exploitation of women and children, etc.?

Again, keep in mind Paasonen's distinction. And, as you make your analysis, be sure to also identify which ethical framework (*utilitarian*, *deontology*, etc.) you primarily draw from.

3. Presuming you agree that a process of pornification is taking place, does it seem to you that the easy access to (as well as, for

example, amateur production of) pornography, especially via such digital media as cellphone cameras, and, of course, the Internet and the Web, has contributed to such a "pornification"?

Be careful! Outside of the scientific laboratory, it's very, very difficult to demonstrate that something in our experiential worlds *causes* something else. Especially in the "culture wars" (in Danish, the *sædelighedsfejde* or "decency controversies") surrounding such things as pornography, precisely a central point of debate is whether or not one social phenomenon (e.g., the legalization of pornography) can be confidently assigned all blame and responsibility for another social phenomenon (e.g., greater violence towards women).

Videogames and virtue ethics

We have seen some (but by no means all) of the arguments that defenders of videogames bring to the table when critics of violent games – perhaps in the midst of a possible moral panic – call for new restrictions. We have also seen that many of the critical arguments stem from either *utilitarian* arguments (e.g., presuming that violent games can be shown to lead to aggression in real life, then such harms must be minimized for the greater good) and/or *deontological* arguments (e.g., if leading to violence, then such games threaten rights to be safe from assault, etc.).

But one of the ethically interesting features of videogames is that some of the prevailing philosophical discussions of them have drawn primarily on *virtue ethics*. In this section, then, we will take advantage of that discussion, and thereby illustrate how virtue ethics, as an ethical framework, may be applied to current ethical dimensions of videogames as an instance of digital media.

A first approach

An initial way of applying a virtue ethics to digital media is to ask the question: what sort of person do I want/need to *become* to be *content* – not simply in the immediate present, but across the course of my entire (I hope, long) life? Along these lines:

what sorts of *habits* should I cultivate in my behaviors that will lead to fostering my reason (both theoretical and practical) and thereby lead to greater harmony in myself and with others, including the larger natural (and, for religious folk, supernatural) orders?

We can refine these questions by looking more carefully at both games in general and then videogames in particular. To begin with, as Miguel Sicart (2005) suggests, games in general bring into play a broad set of ethical guidelines surrounding our notions of *good* sportsmanship: minimally, a *good* player will follow the rules of the game, and accept both winning and losing with a certain amount of grace and respect for the other players. Coupled with our understanding of the game rules, possible strategies, and an awareness of both our own and our opponents' strengths and weaknesses, playing the game further requires us to exercise an Aristotelian sort of *phronesis* or judgment about how to respond best to specific choices in specific situations: our goal in doing so as players is not simply to have fun or to win at all costs, but rather to win (or lose) *fairly* (i.e., by following the rules of the game) and in ways that recognize and respect the abilities of our opponent(s).

In this direction, a virtue ethicist could easily defend videogames (and other forms of play): at least some games provide important, perhaps even essential, sorts of environments and opportunities for effectively developing and improving upon important habits of excellence, including the central capacity of *phronesis* or practical judgment.

At the same time, however, we have seen that the category of computer games covers a wide range of games and interactions – from solitary chess-playing against a computer, through games played online through social networking sites, through serious and educational games, to the sorts of games that tend to dominate especially popular discussion and debate, namely, games involving violence, sex, criminal activities, etc. These last include First-Person Shooter (FPS) games played in solitude on one's own PC or game console, as well as Massive Multiplayer Online Games (MMOGs) such as World of Warcraft.

Initial reflection/discussion/writing questions: don't get violent?

1. If you are a game player, describe the game(s) you are familiar with and play most frequently.

    If you a not a gamer, describe one or more games you've watched others play regularly.

    Either way, what sorts of *habits* or *excellences* are required in order to play these games successfully? That is, what sorts of skills and abilities do they require and foster?

2. Given the habits, skills, etc., that you identify above, can you use one or more of the ethical frameworks we have explored to develop arguments for the playing of such games?

    For example, you might argue from a *utilitarian* framework that playing the game leads to a clear set of *benefits* (e.g., relax-ation, harmless pleasure, improvement of certain skills, etc.) at a modest-to-negligable *cost* (e.g., the cost of the game and required equipment, one's time, etc.)

    Similarly, can you use one or more of the ethical frameworks we have explored to develop arguments *against* the playing of such games?

3. Once you've established – either individually and/or as a group or class – a set of arguments pro and con, how do you respond to the debate here?

    That is, can you develop *additional* arguments, evidence, reasons, etc., that would incline the debate towards one side or another?

4. In the face of these diverse responses and perspectives on the ethical dimensions of computer games, how do you respond?

    In particular: do you respond to these contrasting claims and perspectives as:

    *   an ethical *relativist,*
    *   an ethical *absolutist,*
    *   and/or an ethical *pluralist?*

Explain, and, more importantly, *justify* your response. That is, what additional reasons, evidence, grounds, etc., can you give in support of your *meta-theoretical* response to the first-level debates regarding computer games?

A virtue ethics critique

A virtue ethicist critical of violence in videogames (and other forms of play) might argue that we as human beings should cultivate those feelings and behaviors that lead to greater understanding and harmony between human beings – especially human beings from different cultures, who believe and act in ways sometimes strikingly different from our own. Such cultivation is not always fun: while rewarding in the long run, it is almost always hard work and time-consuming.

By contrast, the pleasure and enjoyment generated by some (but by no means all) computer games – whether in the form of First-Person Shooter (FPS) games played in solitude on one's own PC or game console, or in the form of Massive Multiplayer Online Games (MMOGs) such as World of Warcraft – derive from our focusing on the practices and skills of annihilating others (at least virtually). A critical virtue ethicist might acknowledge that playing games of this sort is ethically justifiable up to a point. But he or she is likely to make at least two arguments against an excessive use of such games. One, as with all our other choices, choosing to spend our time, energy, talents, and abilities in this particular way represents what we might think of as an "opportunity cost": that is, spending two or three hours a day on games of this sort represents the cost of two or three hours possibly spent on other activities – including those activities that might more directly foster the habits and excellences necessary to my becoming a more complete and content human being in community with others. Two, as playing such games reinforces both a basic fear of "the Other" – whether in the form of an alien or an enemy soldier – and my primary response to this Other as one of violence and killing, then it appears that such game playing thereby reinforces beliefs and attitudes that not simply supplant, but arguably work against, the cultivation of other habits and excellences that may well be more important for human development and contentment.

Reflection/discussion/writing: the virtues of videogames?

1. Taking the perspective of a virtue ethicist, can you provide additional argument, evidence, reasons, etc., that either support or contradict these claims as made by a critical virtue ethicist?

     To do so might require, for example, that you consider additional habits of excellence needed for a life of contentment (*eudaimonia*) or happiness (perhaps after reviewing the section on virtue ethics in ch. 6).

     Then your question would be: How far does the consumption and use of videogames that feature violence in the ways described above either foster and/or hinder the pursuit and practice of those habits of excellence?

2. Compare and contrast the approach to videogames outlined above with the approach(es) to videogames you (and/or your cohorts) developed in response to the previous set of questions.

     A. How far do these two sets of arguments "talk to each other" – that is, address the same issues from within a set of shared assumptions and perspectives?
     B. How far do these two sets of arguments "talk past each other" – that is, address the same issues, perhaps, but from different sets of shared assumptions and perspectives?
     C. Insofar as the latter is the case, how far is there a real debate between these two sets of arguments – and how far, in effect, do the arguments run the risk of being irrelevant to the viewpoint of the other perspective, because their basic assumptions are simply not the same?

3. How do you respond to these arguments?

     That is, what in these arguments

(i) works to reinforce and provide additional evidence in favor of your original claims and beliefs regarding violent video-games, and/or

(ii) works to challenge and undermine those original claims and beliefs?

4. Given this more developed perspective, how do you respond to the following claims:

A. Videogames are a harmless pastime, and it is only because "the media" like to focus in on them in trying to explain the acts of a few disturbed people that anyone would believe that violence in videogames is some sort of threat.

B. Violence in videogames demonstrably leads – in at least enough cases to be of legitimate concern to the larger society – to genuine violence and thus harm to others; hence they must be very carefully regulated in order to protect the larger society.

Additional research/reflection/writing questions

1. Mia Consalvo has suggested that when game players ask whether or not a game is "good" or "bad," they have in mind whether or not the game succeeds at entertaining them, if it provides an interesting storyline, etc. By contrast, when politicians and interest groups ask this question, a different set of issues come into play – such as possible glorification of violence (and/or drugs, and/or . . .), how women and minorities are depicted, and, ultimately, what sort of effects exposure to these materials may have on young children. Consalvo goes on to suggest that neither set of questions gets to the genuinely interesting and potentially fruitful ethical dimensions of games – and offers instead her own set of considerations and possible frameworks for beginning to evaluate computer games from an ethical perspective (2005: 11).

Review Consalvo's article, with a view towards developing a summary of her suggestions for how to approach one or more of the ethical questions evoked by computer games.

2. As we might expect, responses to the potential dangers of violent computer games vary from country to country. As part of a special issue on electronic games, authors Dorothy E. Warner and Mike Raiter briefly compare and contrast how different countries have categorized and developed laws around the use of videogames by minors and young adults (2005: 50).

Review their summary and then paraphrase it in your own words as the beginning of a research project/essay on cultural aspects of computer games. In this project/essay, you can pursue through your own research the following questions:

A. What are the prevailing attitudes in your own country/cultural domain regarding computer games, especially if you live in a country/cultural domain *not* described by Warner and Raiter?

In developing your account, you will want to distinguish between the attitudes held by young people vs. older people – most especially, parents and teachers.

B. What are the current laws and guidelines (if any) for the consumption and use of computer games?

C. Insofar as you can uncover *ethical* arguments supporting these laws and guidelines, what *kinds* of ethical arguments are they?

That is, do people appeal primarily to *consequentialist* and *utilitarian* sorts of considerations – e.g., computer games may lead to specific sorts of harms (costs) for both the individual and society, and/or computer games may lead to specific sorts of benefits for both the individual (e.g., learning important skills, especially in playing "serious" or educational games) and society (e.g., economic productivity through their development and sale)?

And/or: do people appeal to other sorts of ethical arguments, such as *deontological* claims (e.g., a *right* to free expression), claims familiar from *virtue ethics* (e.g., framed as concerns regarding the potentially destructive impacts on game players' cognitive and emotive abilities, especially if they are very young), etc.?

3.   Popular media and scientific research: watch your sources!

A.   Popular media – even the most serious and responsible – run the risk of sensationalizing important issues, in order to attract reader interest (and thereby, of course, to increase sales). So, for example, the cover story of the magazine *Der Spiegel* asked the question "Wie viel Computer und Fernsehen verträgt ein Kind? [How Much Computer and TV Can a Child Tolerate?]" underneath a photograph of an expressionless child's face, whose irises were squared off to visually signal the damaging effects of too much time in front of a screen (*Der Speigel 20*, May 14, 2007). The accompanying stories detailed current research into, for example, how excessive TV exposure apparently hinders the cognitive development of very young children ("Geist oder Glotze," 50).

[For the complete article, see Gatterburg 2007. This article can be retrieved for a modest fee from the Spiegel Online Archiv: <http://service.spiegel.de/digas/servlet/find/DID=515364>.]

Using your own research, find two or three examples of how popular media treat one or more of the ethical dimensions of computer games.

B.   If we turn, however, to scientific studies of the emotive and cognitive effects of electronic game playing, the picture becomes enormously complex.

Beginning with the resource list provided below, find and then read and review at least two or three of the publications recommended.

Develop at least brief summaries of these – and then develop a "composite picture" of what this research seems to say regarding one or more ethical dimensions of computer games (e.g., possible harms and benefits, etc.)

C.   Compare and contrast the account of possible harms and benefits of computer game playing as presented in the popular media you researched (A) with the account provided in the scholarly research (B).

Insofar as there are differences between these accounts, what *ethical* conclusions might the popular accounts support, and what *ethical* conclusions might the more scholarly accounts support?

D. If there's a strong contrast between the ethical consequences you can draw from popular accounts vs. scholarly research, what does this contrast itself suggest regarding one's ethical obligation to research carefully into the grounds of one's own claims? That is, is it ethically justified to rely on popular media alone? Do we have any sort of ethical obligation to examine scholarly research that might be relevant to our ethical claims – despite the increased difficulty of doing so?

Additional research resources

Games and the brain[4]

Calvert, Sandra. 2005. Cognitive Effects of Video Games. In Joos Raessen and Jeffrey Goldstein (eds.), *Handbook of Computer Game Studies*, 125–31. Cambridge, MA: MIT Press.
[A largely positive overview of how childrens' videogame play "can cultivate the skills that are necessary for successful navigation along the information highway as well as prepare children for later occupational skills in areas such as engineering and computer programming" (130).]
Lieberman, Debra. 2006. What Can We Learn from Playing Interactive Games? In Peter Vorderer and Jennings Bryant (eds.), *Playing Video Games: Motives, Responses, and Consequences*, 379–97. Mahwah, N.J.: Lawrence Erlbaum Associates.
[An extensive and balanced overview of what can – and cannot – be learned effectively from interactive games, with a particular view towards effective design for specific stages in cognitive and emotional development.]
Weber, René, Ute Ritterfeld, and Klaus Mathiak. 2006. Does Playing Violent Video Games Induce Aggression? Empirical Evidence of a Functional Magnetic Resonance Imaging Study. *Media Psychology 8*, 39–60.
[Uses functional magnetic resonance imaging to determine the possible effects of playing violent videogames, showing that "virtual

---

4 I wish to express my gratitude here to Drs. Debra Lieberman and René Weber, Department of Communication, University of California-Santa Barbara, for providing this list of suggested resources.

violence in videogame playing results in those neural patterns that are
considered characteristic for aggressive cognition and behavior"
(51), but that "even if the neural patterns in real life and virtual
experiences are identical, the experiences may not be. Thus, the
equivalence of subjective experiences and brain activity is
questionable" (53).]

More on games

Bogost, Ian. 2007. *Persuasive Games: The Expressive Power of Videogames.*
Cambridge, MA: MIT Press.

Consalvo, Mia. 2007. *Cheating: Gaining Advantage in Videogames.*
Cambridge, Mass: MIT Press.

Gee, James Paul. 2003. *What Video Games Have to Teach Us about Learning
and Literacy.* New York: Palgrave Macmillan.

Sicart, Miguel. 2009. *The Ethics of Computer Games.* Cambridge, MA: MIT
Press.

Wonderly, Monique. 2008. A Humean Approach to Assessing the Moral
Significance of Ultra-Violent Video Games. *Ethics and Information
Technology 10* (1), 1–10.

Pornography and all that . . .

Barber, Trudy. 2004. A Pleasure Prophecy: Predictions for the Sex Tourist
of the Future. In Dennis D. Waskul (ed.), *net.seXXX: Readings on Sex,
Pornography, and the Internet*, 322–36. New York: Peter Lang.
[An extensive set of predictions as to the possibilities of virtual sex –
useful primarily for suggesting possible scenarios for further
discussion of the ethical dimensions of virtual sex.]

Durkin, Keith F. 2004. The Internet as a Milieu for the Management of a
Stigmatized Sexual Identity. In Dennis D. Waskul (ed.), *net.seXXX:
Readings on Sex, Pornography, and the Internet*, 131–47. New York:
Peter Lang.

Hughes, Donna. 2004. The Use of New Communications and
Information Technologies for Sexual Exploitation of Women and
Children. In Dennis D. Waskul (ed.), *net.seXXX: Readings on Sex,
Pornography, and the Internet*, 109–30. New York: Peter Lang.

Mowlabocus, Sharif. 2007. Gay Men and the Pornification of Everyday
Life. In Susanna Paasonen, Kaarina Nikunen, and Laura Saarenmaa
(eds.), *Pornification: Sex and Sexuality in Media Culture*, 61–71. Oxford,
New York: Berg Publishers.

[A study of a U.K. gay website, finding that, contrary to the potential of online venues and pornography to expand the possibilities of sexual identity, these rather work to reinforce a relatively narrow sexual ideal.]

Nikunen, Kaarina. 2007. *Cosmo* Girls Talk: Blurring Boundaries of Porn and Sex. In Susanna Paasonen, Kaarina Nikunen, and Laura Saarenmaa (eds.), *Pornification: Sex and Sexuality in Media Culture*, 73–85. Oxford, New York: Berg Publishers.
[An exploration of how far pornography shapes young girls' understandings of sexuality, etc., based on an analysis of a discussion board affiliated with *Cosmopolitan* magazine.]

Rheingold, Howard. [1990] 2004. Teledildonics: Reach Out and Touch Someone. *Mondo 2000*. Issue 2/Summer: 52–4, reprinted in Dennis D. Waskul (ed.), *net.seXXX: Readings on Sex, Pornography, and the Internet*, 319–21. New York: Peter Lang.

Roberds, Stephen C. 2004. Technology, Obscenity, and the Law: A History of Recent Efforts to Regulate Pornography on the Internet. In Dennis D. Waskul (ed.), *net.seXXX: Readings on Sex, Pornography, and the Internet*, 295–316. New York: Peter Lang.
[A good overview of the relevant history, but focused almost exclusively on the United States context.]

Stuart, Susan. 2008. From Agency to Apperception: Through Kinaesthesia to Cognition and Creation. *Ethics and Information Technology* 10 (4), 255–64.
[Drawing in part from cognitive psychology and neuroscience, Stuart provides an important overview of contemporary understandings of the role of the *body* in how we know the world and navigate through it – thereby countering a number of 1990s' claims about the possibility and desirability of migrating effortlessly into cyberspace. With this as background, Stuart introduces the possibility of a "virtual reality adultery suit" to highlight the ethical issues evoked by a blurring of traditional boundaries between the virtual and the real. A helpful counterbalance to Thomas 2004.]

Thomas, Jim. 2004. Cyberpoaching behind the Keyboard: Uncoupling the Ethics of "Virtual Infidelity." In Dennis D. Waskul (ed.), *net.seXXX: Readings on Sex, Pornography, and the Internet*, 149–77. New York: Peter Lang.

White, Amy. 2006. *Virtually Obscene: The Case for an Uncensored Internet*. Jefferson: McFarland & Company.
[White critiques many of the common arguments for censoring pornography online, beginning with the claims regarding its harms – primarily, of women and children. She further argues against

censorship – primarily on the grounds that it can be a slippery slope, leading to censorship of other materials, and that such censorship, at least on the Internet, is technologically very difficult.

White further raises the central issue that in the U.S., anti-obscenity statutes are based on local community standards – i.e., what a local community may find offensive, obscene, etc., defines obscenity in that community. But given the globally distributed nature of the Internet, how can such statutes be plausibly applied?]

CHAPTER 6

Digital Media Ethics

Overview, Frameworks, Resources

Morally as well as physically, there is only one world, and we all have to live in it.

(Midgley [1981] 1996: 119)

Chapter overview

This chapter is intended to provide especially new students of ethics with an overview of some of the most commonly used *theoretical* frameworks for ethical analysis and decision-making. In this direction, we begin with discussion (and writing exercises) regarding (1) *utilitarianism* and (2) *deontology*. Before turning to additional theoretical frameworks, we then explore (3) important *meta-theoretical* frameworks – frameworks for thinking *about* the theoretical frameworks. (As we will see, the diversity of theoretical frameworks naturally raises questions such as: which ones – if any – are right? And, can more than one be right? And so forth. These sorts of questions are thus the focus of the meta-theoretical frameworks of *ethical relativism, ethical absolutism* [*monism*], and *ethical pluralism*.) With these meta-theoretical frameworks in hand, we then return to additional theoretical frameworks: (4) *feminist ethics and ethics of care*, (5) *virtue ethics*, (6) *Confucian ethics*, and (7) *additional perspectives – African*.

Taken together, these theoretical and meta-theoretical frameworks constitute a kind of "ethical toolkit" – a collection of important but diverse ways of analyzing and attempting to resolve ethical problems. As we will see, part of our work as ethicists is not simply to learn how to apply a given *theoretical* framework to a specific issue; given the diversity of possible theoretical frameworks, we must also make (it is hoped informed and well-reasoned) choices regarding which frameworks are best suited for confronting and resolving our ethical issues – choices guided in turn by the *meta-theoretical* frameworks of relativism, absolutism, and pluralism.

A synopsis of digital media ethics

Not surprisingly, much of the ethical reflection on digital media – most especially, on the ethical dimensions of information and communication technologies (ICTs) – arose alongside the technologies themselves. But this means that until very recently, most of the discussion and reflection on digital media ethics took place primarily within Western countries, and thereby brought into play primarily Western ethical traditions and ways of thinking.

So, for example, Terrell Ward Bynum points out that the first book on computer ethics was written by Norbert Wiener in 1950 (*The Human Use of Human Beings: Cybernetics and Society* – see Bynum 2000). For over two decades, "computer ethics" was the concern of a very small group of professionals – primarily computer scientists and a few philosophers. "Computer ethics" as its own term, in fact, emerged only in the 1970s, primarily through the work of Walter Maner, but also manifest, for example, in the first professional code of computer ethics as established by the Association for Computing Machinery in 1973 (and subsequently revised – most recently, in 1992). The introduction of the personal computer (PC) in 1982, however, began a dramatic expansion of the role of computers and computer networks well into the lives of "the rest of us" – i.e., those of us who are *not* computer scientists or other sorts of information professionals such as librarians (see Buchanan and Henderson 2008). Following the emergence of the Internet and World Wide Web into the lives and awareness of most people in the developed world in the early 1990s, a number of savvy observers began (rightly) to predict that information and computing ethics (ICE) would become a *global* ethics by the beginning of the twenty-first century – i.e., a domain of ethical issues, debate, and possible resolution of concern to more and more people representing an increasingly global diversity of cultural norms and ethical and religious traditions (see Paterson 2007: 153). And in fact, in the domain of ICE, what is called "intercultural computing ethics" has been underway since the 1990s (Capurro 2005, 2008; Ess 2005).

Along the way, an important *meta-ethical* debate has emerged. As we have seen, digital media present us with at least three sorts of ethical challenges. First, they raise for us ethical problems already *familiar* from our use of more traditional media – e.g., whether or not to illegally copy a song for one's own personal use. But, second, these familiar difficulties are now sometimes accompanied by new wrinkles – e.g., the *ease* with which such copies can be (perfectly) made, cheaply stored, and rapidly distributed makes such copying all the more tempting than in the days of vinyl records and reel-to-reel tape machines. Third, new media may present us with distinctively *new* ethical problems. Perhaps most dramatically, as persons in important ways are ever more dependent on *information* about them (including personal identification records, banking and other financial transactions, etc.), it is by no means clear that "privacy" as understood in the modern but pre-digital era can be meaningfully sustained. And, as we saw in chapter 4, our being globally interconnected with one another may mean that all of us, not just a few, are obliged to learn the virtues of cosmopolitan communication.

Hence a debate emerged in computer ethics as to whether or not ICE, especially as it becomes globalized, will represent (a) largely a continuation of traditional ethics, but now applied to new problems, or (b) a radical transformation of ethical thinking, as ICTs introduce us to radically new possibilities of interaction with one another, such that the ethical difficulties affiliated with these new technologies can no longer be resolved using traditional ethical frameworks and approaches (see Bynum 2000; Tavani 2007: 12).

For our purposes, the important point is to be aware of this larger meta-ethical question and debate as we go along. Our reflections and responses to this question will affect (and be affected by) our ethical reflection regarding other digital media.

(From my perspective, ethical reflection on other digital media – e.g., cameras, copiers, cellphones, voice recorders, etc. – has largely

taken place on the assumption that such media represent exten-
sions of traditional media, and, thereby, the ethical issues they raise
can be analyzed and resolved in more or less the same ways.
Whether or not this is the case will be an issue for further discussion, reflection, and research as we go along.)

Basic ethical frameworks

As we have seen in the opening chapter, "doing ethics" involves
much more than simply picking a set of principles, values, etc., and
then applying these in a largely deductive, algorithmic manner to a
problem at hand. Rather, our central ethical difficulties are difficult
in large measure because they require us to *first* determine which
principles, values, frameworks, etc., in fact apply to a given
problem – a determination that Aristotle famously associated with
the capacity for *practical judgment* or *phronesis*. Developing such
judgment requires nothing less than an on-going effort to analyze
and reflect on both familiar and new experiences and problems.
The good news is that our ethical judgments – at least, if we consciously seek to develop them in these ways – generally do get
better over time. The somewhat daunting news is that developing
such judgment is a life-time's work, one that is in some important
sense never complete or final.

Nonetheless, we must start somewhere. (As we've seen Socrates
say in *The Republic*, whether we are in a swimming pool or an
ocean, we must start swimming all the same.) The following material is intended to introduce you to some of the most basic and
widely used frameworks for ethical reflection – beginning with
characteristically Western ones, but then moving on to non-
Western ones as well.

Reflection/discussion exercise: a student dilemma

It's Wednesday evening, and you're packing up some books and
notes to take over to a friend's apartment. You have different
majors, but you are both in the same section of a required
course – and tomorrow is one of two exams given during the

semester; your grade on the exam will count towards 40% of your final grade in the course.

For you, the course is not so hard, but your friend is really struggling. You've promised to help her study this evening; you both need to get a good grade on the exam and in the course to keep your grade point average at the level required for your scholarships.

Just as you're walking out the door to go to your friend's apartment, a good friend calls you up and says that he and some of your buddies are at the local pizza place, having dinner and some beers. They'd really like you to come on over – in part, because you owe them a round or two of drinks from the last time you got together.

What do you do?

1. Utilitarianism

Most students in my experience approach this sort of problem in a *consequentialist* – perhaps even a *utilitarian* – way. That is, they will begin to figure out: what are the *costs* and *benefits* of (1) turning down their buddies for pizza and beer vs. the costs and benefits of (2) fulfilling the promise to help a friend study.

One of the chief advantages of this approach is that we can set up a handy table to help us keep track of the positives and negatives. An initial analysis of our choices might look like the table on p. 172.

But of course, there are additional positive and negative consequences of our choices that may seem relevant to our decision: e.g., if I help my friend, she will do better on her exam (and, most likely, so will I); if I go have pizza and beer, I will certainly have a good time this evening, but probably not do so well tomorrow on the exam. If we think further down the road, it may be that doing well on this exam will turn out to be a "make-or-break" event with regard to our success in the course: that is, should we both do well, we might subsequently end up with a better grade in the course; but if we don't, then we might end up with less of a grade than we need in order to maintain our grade point averages for our scholarships, etc.

Consequentialist analysis	Possible actions	
	Fulfill promise – study with friend	Break promise – enjoy pizza and beer
Costs (negatives)	Will miss a nice evening with friends	Will disappoint a friend who's counting on your help . . . . . .
Benefits (positives)	Will be able to help a friend in an important way	Will enjoy a nice evening with friends . . . . . .

The possible consequences even further down the road might be enormous – ranging from doing well in school more generally, moving on to a good job, etc., to (worst-case scenario) losing needed scholarships, thereby being unable to complete school, thereby failing to be able to find a good and satisfying job, etc.

You get the point. For the *consequentialist*, the game of ethics is about trying to think through possible good and bad consequences of possible acts, and then weighing them against one another to determine which act will generate the more positive outcome(s).

Strengths and limits

Consequentialism is certainly a tried-and-true approach to ethics: it's at least as old as Crito's efforts in the dialogue named after him to persuade Socrates to break out of jail and thereby avoid execution by the Athenians. And especially in its *utilitarian* form – i.e., as developed in the modern era by Jeremy Bentham and further elaborated by John Stuart Mill, both arguing that we must pursue those acts that bring about the greatest positive consequences (pleasure) for the greatest number – such consequentialist approaches have come to dominate ethical decision-making, especially in the United States and the United Kingdom (or so some analysis suggests).

Certainly, there are many cases in which consequentialism will do what we want an ethical theory to do – i.e., to help us determine which is the better choice of two (or more) possible actions.

But as this example also suggests, consequentialist approaches face serious limitations. (We will also see this to be true of every other theory we examine: after we have reviewed all the theories under discussion here, one of our questions will be to see if we can discern which theory – or, perhaps, a combination of theories – seems more sound, useful, justifiable, etc., than its competitors.) In my view, there are three important such limitations.

(a) *How do we evaluate [e-valuate = place value on] the possible conse-quences of our acts?* In simple cases, this is not a problem. Either I go get a new bus pass or I face walking to school on a cold winter day. Either I pay my phone bill or find myself out of touch with friends and family trying to call me and send me text-messages, etc.

But the hard cases are hard in part as it's not always clear how we are to *weigh* the possible outcomes of one act against another.

Bentham famously thought that all possible consequences, as some form of pleasure or pain, could thus be ultimately measured and given a value – a positive or negative "util." Ethical decision-making would then be a strictly computational matter of adding up positive and negative utils.

But what if everything cannot be measured solely in terms of pleasure or pain? What number of utils do we assign to an evening with friends, enhanced by the pleasures of food and drink? What number of utils do we assign to breaking a promise to a friend, coupled with the knowledge that our breaking that promise may lead to further, perhaps very serious consequences (= negative utils) for our friend?

As we will see shortly, *deontologists* believe that there are some aspects of human existence that can*not* be assigned quantitative values. Hence both consequentialist approaches in general and utilitarianism in particular have no ethical legs to stand on: without a universal and consistent schema of positive and negative utils with which to make our calculations, the calculations at the heart of consequentialism cannot proceed. Moreover, in this case, for the deontologist, a promise is a promise; it thereby has an absolute (or near-absolute) quality that means that to break a promise, even though the promise-breaker might get great pleasure as a result of doing so (i.e., because doing so will open the door to pizza and beer), is still wrong.

But we don't have to be deontologists to recognize that there's a problem here: everything turns in consequentialism on assigning relative weights or values to given consequences, and it's simply not always clear how we are to do so.

(b) How far into the future must we consider? Ethicists helpfully distinguish between short-term and long-term consequentialists. In this example, a (really) short-term consequentialist would consider only the consequences of his or her acts over the next few hours. For most of us – at least, if we're not allergic to gluten and if our religion or physiology does not forbid alcohol – pizza and beer with friends would generate more positive utils than studying for an exam (presuming, that is, that you really do *not* like the subject, etc.). By contrast, extending our timeframe by 24 hours might radically change our decision: whatever the positive utils of pizza and beer, they might well not outweigh the negative utils of letting down a friend and then watching as both of us do poorly on an important exam.

And so on. It's not inconceivable that in twenty or thirty years you and your friend might look back on this exam as a key moment in your lives – one that led (in the best of circumstances) to further academic and thereby vocational success or (perish the thought) to academic failure and a lifetime of mediocre and unsatisfying jobs. The difficulty is: consequentialists and utilitarians do not appear to have a satisfying justification for telling us *where* in time to draw the line – the point after which we no longer need worry about the outcomes of our choices. But depending on where we draw this line can make all the difference in our calculations.

As this last point suggests, there's actually a second difficulty wrapped into the question of how far into the future must we attempt to consider: pretty clearly, the further out into the future we seek to predict, the less reliable our predictions can be. And yet, some of those future consequences may be some of the most important for us in our lives. Worst-case: the chances of realizing what may potentially be the most decisive consequences of our acts become increasingly (perhaps vanishingly) small the further into the future we seek to predict those consequences.

(In my experience, much of the anguish we face in ethical decisions turns on our effort to approach them in a consequentialist fashion – only to realize that we cannot be very certain at all about some of the most important possible outcomes of our actions.)

(c) For whom *are the consequences that we must consider?* The pizza-and-beer example takes into account only a small number of people. Bentham and Mill, by contrast, were far more ambitious, thinking that consequentialism would work as applied to whole societies. Up to a point, at least, this seems to be plausible. Especially in wartime, for example, generals and political leaders think in clearly consequentialist terms. Deciding to drop the atomic bombs on Hiroshima and Nagasaki, for example, were relatively easy decisions for the Allied commanders. Dropping these bombs immediately cost something like 200,000 Japanese deaths – but, as hoped, it put an end to the war. A conventional land invasion was estimated to result in ca. 500,000 Allied soldiers' deaths (and at least as many Japanese soldiers). At a simple assignment of one positive util per life:

- to use atomic weapons: 500,000 + / 200,000 − = 300,000 + utils
- not to use atomic weapons: 200,000 + / 500,000 − = 300,000 − utils

But what about the impact of using these weapons on those who continued to live (and die) in areas contaminated by radioactive fallout? What about the impact of using these weapons on the larger ecosystem? On future generations?

Attempting to take these possible consequences into account clearly makes the calculation much, much more complicated. Again, part of the problem is attempting to determine how far into the future we must predict relevant consequences. But the further problem is: where do I draw the line with regard to consequences affecting what *group* of persons/living beings/non-animate entities? As I hope is clear, depending on where I draw that line can make an enormous difference in the possible consequences of an act – and, thereby, how I decide which of two (or more) competing choices I should pursue.

In particular, as digital media radically extend the range of the possible consequences of our actions (as dramatically illustrated in the example of the Muhammad cartoons), the question of "consequences for whom" becomes central. Unlike commanders in war, we cannot simply assume that the consequences of our actions are limited to the citizens of a given nation-state.

In the face of these sorts of difficulties and limitations, many people find that they cannot rely on consequentialism alone. They may want to retain consequentialist approaches for certain sorts of decisions – e.g., when it is possible to make reasonably reliable predictions about the possible outcomes of our choices; when it is reasonably clear who will be affected, and within a specified timeframe. But especially when this sort of insight and information are not available, they may turn to one or more of the following ethical frameworks.

2. Deontology

For deontologists, what stands out in our opening example is that you have made a promise. And for deontologists, promises – along with, say, notions of basic rights and duties – have an absolute quality to them, one that means that they can*not* be overridden by such considerations as to how much pleasure (or pain) might be gained (or avoided) by violating them.

Religiously grounded forms of deontology are perhaps most immediately familiar to contemporary Westerners. For example, if I am a Jew, Christian, or Muslim, I believe that God has given us specific commandments and laws which define right and wrong for me – no matter the consequences. So, negatively, I am commanded not to murder, not to lie, not to covet my neighbor's property, not to commit adultery, etc. Positively, I am commanded to love God and neighbor; indeed, the Golden Rule is a commandment found in every major religious tradition around the world. Hence, a religiously grounded deontologist would believe that it is wrong to lie – even if, by lying, he or she might be able to gain significant material reward.

As a still stronger example: religious pacifists – whether rooted in Judaism and Christianity or, for example, in some forms of Buddhism – take the sacredness of life (all life for the Buddhist, not just human life) as an absolute. Hence, for pacifists, killing other human beings (and, for many Buddhists, any living thing) is always wrong – no matter the consequences. Such pacifists would not only reject the consequentialist thinking, for example, behind the decision to use atomic weapons in World War II; they further reject the use of violence against others even in self-defense. This is to say: for the religious pacifist, killing another is *always* wrong, no matter the consequences – including the possible consequence of losing one's own life.

[Such pacifism, we can note, can also be supported by consequentialist considerations. Socrates, for example, argues in *The Republic* and *The Crito* that doing violence or harm to another has ultimately unacceptable consequences. Such harm is understood as working contrary to the central ability of reason to discern the good and the ability of judgment (*phronesis*) to determine how to enact the good properly in specific contexts and circumstances. To work contrary to these functions of reason and judgment in turn runs the risk of degrading – perhaps ultimately paralyzing or destroying – these central abilities. And if we degrade or destroy our ability to discern the good and judge what it means, we thereby will lose our ability to make the judgments needed to pursue a genuinely good life of contentment (*eudaimonia*) and harmony. Failure to achieve these, finally, makes our lives no longer worth living. Hence, the just or good person will never harm another, no matter what sorts of other gains such harm might bring, because to do so risks making life no longer worth living (e.g., *The Republic* 335b–335e). However we ethically understand the pacifism of Jesus and the early Christian communities, Gandhi and Martin Luther King, Jr., famously built on these Socratic and Christian roots (and for Gandhi, at least, the Buddhist virtue of *ahimsa*, nonviolence) to argue and *practice* nonviolent protest against unjust laws. Such nonviolence was intended to not only prevent harm to the selves or souls of its practitioners (one consequence), but also to awaken the conscience of the larger community (a second consequence), in hopes that the larger community would come to see the injustice of its behaviors, laws, etc. (consequence 3) and then replace these with more just ones (consequence 4).]

But there are also *rationalist* deontologies – perhaps most importantly, the deontological ethics articulated in the modern era by Immanual Kant (1724–1804). Kant is famous for developing what he called the Categorical Imperative – which we can think of as a kind of *procedural* way of determining what actions are right. Briefly (and, probably, too simply), the first formulation of the Categorical Imperative states: "So act that the maxim of your will could always hold at the same time as a principle establishing universal law" ([1788] 1956: 31). We can see what this means in an initial way by using one of Kant's own examples from *The Foundations of the Metaphysics of Morals* ([1785] 1959). Consider the possibility of needing to borrow money, knowing full well, however, that you will not be able to repay the loan. You also know that in order to get the loan, you have to promise to repay it, of course. Question: can you make what you know to be a false promise in order to secure the loan? For Kant, the *maxim* of this action would be: "When I believe myself to be in need of money, I will borrow money and promise to repay it, although I know I shall never do so" ([1785] 1959: 40). But the Categorical Imperative requires that we ask: "How would it be if my maxim became a universal law?" ([1785] 1959: 40).

This is perhaps recognizably close to your parents' asking you in high school: what if everyone did that? But for Kant, what is at stake in this question is whether or not the larger social order that would result from *everyone* following the maxim of "make a false promise when it is convenient to do so" would be coherent – or logically contradictory. On Kant's analysis, attempting to universalize this maxim would become self-contradictory in a critical way:

> For the universality of a law which says that anyone who believes himself to be in need could promise what he pleased with the intention of not fulfilling it would make the promise itself and the end to be accomplished by it impossible; no one would believe what was promised to him but would only laugh at any such assertion as vain pretense. ([1785] 1959: 40)

That is, if we knew that everyone would lie when convenient (the result of universalizing the maxim of our action), then we would never know when someone was telling us the truth. But a world in which we by default cannot trust one another to make promises in good faith – i.e., to tell the truth when we promise one another, for example, to repay a loan – would be a world in which promises would thus lose their meaning. Specifically, in this case, the attempt to lie in order to acquire a loan I have no intention of repaying becomes self-contradictory: if everyone else allows himself or herself the same act – which would result from universalizing the maxim at work here – then no one would accept my promise at the outset. But if I cannot universalize lying in this way – i.e., make it a universal law acceptable for everyone – then for Kant it is wrong, even when it seems convenient or important. That is, it is always wrong, no matter the consequences.

In our case, a Kantian analysis would be: ask the question, what sort of social/moral order would result if everyone were to break a promise whenever doing so would result in at least more immediate, short-term pleasure? It seems likely that we would never be able to trust anyone's promise – which would make promise-making self-contradictory and meaningless. Hence, breaking a promise is always wrong – no matter the consequences.

Finally, deontology is apparent in the widely shared belief that there are ethical absolutes such as *human rights*. The discussion and literature on rights is largely modern: so Thomas Jefferson, inspired by John Locke, insisted in the *Declaration of Independence* that

> We hold these truths to be self-evident: that all men are created equal; that they are endowed by their Creator with inherent and unalienable Rights; that among these are Life, Liberty and the pursuit of Happiness . . . . ([1776] 1984: 19)

The belief in human rights inspired the American and French revolutions – and, subsequently, much of the political transformations that define modern Western states.

But the belief that rights exist as absolutes that must be recognized and protected is not simply a Western phenomenon. By 1948, the United Nations issued its Universal Declaration of Human Rights – a document that goes well beyond what some scholars call the first-generation or primarily *negative* set of rights articulated by Locke and Jefferson, to include second-generation or *positive* rights – e.g., the rights to education and health care. These rights have been realized, for example, as duties of the state in Western Europe and Scandinavia, while the right to health care remains hotly disputed in the United States. For that, a *deontological* notion of basic human rights has driven much of the political activism and transformation of modernity, both within and beyond the boundaries of "the West."

We will see that the claims that human rights exist as universal values will lead to important questions regarding the role of *culture* in shaping our ethics – questions that become especially pressing as digital media increasingly make it possible for individuals from diverse cultures around the globe to communicate quickly and easily with one another.

Difficulties . . .

As these examples suggest, deontology is open to criticism precisely because of its absolute nature.

That is, on the one hand, we may agree, for example, that consequentialism becomes suspect when it leads us to violate what we may take to be absolute human rights. That is, using the utilitarian mantra of "the greatest good for the greatest number," we might argue that the sacrifice of the few for the good of the many is justifiable. We certainly make this argument in wartime, when soldiers, by definition, are those whose lives are potential sacrifices for the good of the many. But these days, we may be less sympathetic to similar arguments that could be made, for example, regarding enslavement. That is, a utilitarian might argue that just as it would be ethically justified to sacrifice a comparatively small portion of the population (soldiers) for the sake of the greater good, so we can justify the loss of certain freedoms and rights of a few (slaves) if we can show that these costs are overridden by the greater benefits such slaves would provide for the larger society. If we wish to argue *against* the utilitarian at this point, we may do so by reaching for some notion of (near-)absolute human rights – e.g., rights to life, liberty, and the pursuit of property. If, as modern deontologists would argue, these rights exist and are (near-)absolute, then they may never be violated – e.g., by turning some portion of the population into slaves – even if to do so might lead to greater pleasure and enjoyment on the part of everyone else.

Likewise, we might admire the courage of protestors, e.g., during the civil rights movement of the 1960s or in more recent political protest movements around the world, who practice the non-violent pacifism of a Gandhi or a King, sometimes with remarkable success. If we are deontologists, we would say that they are doing the right thing – even if it costs them great personal pain, and even if they are not always successful in gaining their intended political outcomes.

But many people are not always willing to accept a Kant-like absolute not to lie, for example. Sometimes, it seems quite clear that lying would be justified – e.g., if it were to save a life, much less many lives.

Indeed, to be fair to Kant, he developed a more nuanced position in his later works, so as to make greater ethical room for deception: while we might deceive others for less than ideal reasons, as deception allows us to hide our more negative characteristics while nonetheless developing a more virtuous character, it can help us become better persons. (See Myskja 2008 for discussion and application to the question of how we develop *trust* online.) As Kant's own transformation suggests, whether or not deontological approaches can consistently make room for what appear to be justified and important "exceptions to the rule" is a central question for defenders of this approach.

Discussion/reflection/writing: a first go at ethical theory

1. Given this initial overview of consequentialist vs. deontological approaches, review the initial example of promise-keeping vs. enjoying pizza and beer with friends. In particular:

A. How did you initially analyze the dilemma – i.e., more as a con-
    sequentialist, and/or more as a deontologist? (As the use of the
    "and/or" suggests, while there are sharp differences between
    the two positions, it is possible for us to use both in some
    combination or another.)

B. Now that you've had a chance to review and explore these two
    frameworks, try applying them to another ethical dilemma –
    ideally, one affiliated with the use of digital media.
    In doing so,

    (i) Describe the dilemma as fully and accurately as you can.
    (ii) Explain what your own initial response to this dilemma
         might be. That is, *what* would you decide to do, and *how*
         would you decide what to do?
    (iii) Then apply each of these frameworks to the dilemma as
         best you can – perhaps with the help of cohorts and/or your
         instructor. Make clear how each framework leads to a given
         outcome or decision regarding possible acts or choices.

C. Given the dilemma you choose to analyze, does the consequen-
    tialist approach lead to the same ethical conclusion as a
    deontological approach, or to a different one?
    Especially if the outcomes are *different*, which outcome more
    closely fits with your own initial response to this dilemma (i.e.,
    your response in B.ii)?

D. Especially if your initial response(s) meshes well with either a
    consequentialist and/or deontological response, do you see any
    additional reasons, insights, arguments, analytical approaches,
    etc., offered by either consequentialism and/or deontology
    beyond those that you initially used in approaching this problem?
    What are these – and do you think that these may prove
    useful in approaching other ethical dilemmas as well? (In
    Kantian terms: can you universalize these – or are they just
    useful in this particular case?)

E. Especially if the outcomes of these two different approaches are
    different, what does this difference *mean*? That is, are we forced,

for example, to choose between one or the other approach, such that one is always right and the other always wrong?

If you say "yes," can you justify (provide good reasons, argument, evidence for) your response?

If you say "no" – again, can you justify (provide good reasons, argument, evidence for) your response?

3. Meta-ethical frameworks: relativism, absolutism (monism), pluralism

Ethical relativism

These contrasts between utilitarian and deontological ethics suggest, on first glance, a *meta-ethical* view called *ethical relativism*. That is, in the face of (often radically) different ethical frameworks and claims, it is tempting to believe that these differences can only mean that *there are no universally valid ethical norms, values, approaches*, etc. Rather, it is argued, all such norms, values, and approaches are valid only *relative* to (i.e., within the domain of) a given culture or group of people. Such ethical relativism is even more tempting as we gain more knowledge and experience of how people live, think, and feel in cultures different from our own – a knowledge increasingly easy to acquire in a world increasingly interconnected by digital media.

Ethical relativism offers us two chief advantages. First, it allows us to *tolerate* the views and practices of "Others" – i.e., those who are different from ourselves. Such toleration is itself an important ethical value; generally, it seems, the world could do with much more tolerance of important ethical (and cultural) differences, not less. Second, ethical relativism offers a certain kind of relief: if values and practices are always and only legitimate in relation to a specific culture, then we are under no obligation to look further for values, practices, frameworks, etc., that might claim genuinely *universal* validity. This latter task is indeed very hard work – a task that is perhaps too much to ask of all but professional philosophers and ethicists. Ethical relativism gives us the excuse and rationale we need to dismiss this task.

In my view, ethical relativism enjoys a third virtue: in at least some instances, it seems to be *true*. For example, in Germanic countries, there is a strong obligation to show respect for one's cohorts at a party by shaking hands with each of them (i.e., not only the hosts, but also all the guests) before leaving. In the United States, there is no such compunction. On the other hand, it is very common for U.S. persons to hug one another when greeting – including (at least in my Midwestern university) colleagues. Doing so in a Germanic culture, by contrast, is almost never appropriate. At least on first glance, it appears that there is no absolute right or wrong regarding such greeting rituals. Rather, what is right in Germanic cultures would seem strange in a U.S. context, and what is right in the U.S. would seem very strange indeed in Germanic cultures.

(That said, we will see in the discussion of ethical *pluralism* that these differences in greeting rituals may not be quite so absolutely relative as they first appear; but for now, it seems that we can safely say that what is right in one cultural domain is wrong in another, and vice versa.)

But ethical relativism also faces two especially important *difficulties*. First, it is logically incoherent – and this in two ways. To begin with, the ethical relativist faces a simple, but fundamental contradiction: on the one hand, s/he wants to argue that there are no universally valid values, norms, practices, etc.; on the other hand, s/he concludes that we must thereby be *tolerant* of peoples and cultures whose ethical norms and practices may be different from our own. (Just to be clear: we can get to this tolerance in other ways, as we will see below in the section on ethical pluralism.) But *tolerance* thereby appears to emerge as itself a universally valid ethical norm or value – i.e., one that the ethical relativist wants to say we *all should* agree upon and follow.

Hence, the position of ethical relativism seems caught in a fundamental contradiction: if all ethical values, norms, and practices are indeed valid or legitimate only in relation to a given culture or time, then it would seem that tolerance must likewise count as only a relative value. And so if there are those who are rigidly intolerant on some point – e.g., the eighteenth-century white racist's intolerance for people of color – it is not at all clear how the ethical relativist can coherently insist that such a person, as a product of a given culture and time, should rather have exercised tolerance instead.

The second logical problem for the ethical relativist is somewhat more difficult to show and understand – but let's try. The primary argument for ethical relativism can be put as follows:

- (Premise 1): If there are no universally valid values, practices, beliefs, etc., then we would expect to find diverse ethical values, practices, beliefs, etc., in diverse cultures and times.
- (Premise 2): We *do* find diverse ethical values, practices, beliefs, etc., in diverse cultures and times.
- (Conclusion): Therefore, there are no universally valid values, practices, beliefs, etc.

In logical terms, this argument commits the basic fallacy of *affirming the consequent*.

To see that this argument is a fallacy, consider another argument that uses the same *form* as this one:

- If you like strawberry-flavored gum, then we can expect to find a red-colored packet of gum in your pocket.
- (Premise 2): We *do* find a red-colored packet of gum in your pocket.
- (Conclusion): Therefore, you must like strawberry-flavored gum.

While this seems sensible enough, it takes only a little reflection to see that both the first and second premises could be true, but the conclusion must *not* be true: perhaps you've switched to cinnamon-flavored gum today, which also comes in a red-colored packet?

To return to the original argument: the argument is a fallacy – meaning, the conclusion does *not* necessary follow – because it is *possible* that we find diverse values, beliefs, practices, etc., in diverse cultures *for other reasons* besides the one offered in the first premise (i.e., that there are no universally valid values, beliefs, practices, etc.). As we will explore more fully below, the meta-ethical position of *ethical pluralism* argues precisely that these diverse values, beliefs, practices, etc., are the result of diverse *interpretations/ applications/understandings* of *shared* ethical norms.

The debate between ethical relativists and ethical pluralists is an open, on-going, and important one – one that we will further reflect upon in additional reflection and writing questions. But at this juncture, the crucial point is this: *if* there are plausible *alternative* reasons for our observing diverse practices, beliefs, norms, etc., than just the one claimed by the ethical relativist (i.e., that there are no universally valid norms in the first place), then the argument for relativism is not a valid one.

The second set of objections against ethical relativism center on the arguments that seek to show that ethical relativism can actually work *against* the sort of tolerance and mutual understanding that it seems to endorse and that makes it so attractive. Again, this involves two elements. To begin with, ethical relativism – by design – will not let us make any sort of ethical judgment about "the Other," i.e., the person whose values, beliefs, practices, etc., are different from our own because, it is argued, they are the product of a different culture, time, etc. But this means, for example, that those raised in the United States and the United Kingdom can neither praise a Mother Teresa in India as a moral hero, nor condemn a Hitler in Germany as a moral monster. In this way, ethical relativism leads to a paralysis of moral judgment – a paralysis that would require us to accept, say, genocide in Rwanda, rape-rooms and rape as terror in war, the use of babies and children as carriers of explosives in suicide bombings, etc., etc. (Indeed, ethical relativism could be used to paralyze moral judgment within one's own backyard: systematic violence against women, for example, could be excused as part of the "culture" of a given religious group, even if I personally am appalled by it.)

Moreover, Mary Midgley ([1981] 1996) has argued that ethical relativism further leads to what she calls moral isolationism. On this view, we presume that there is a water-tight boundary between specific cultures. This boundary not only prevents us from making ethical judgments about the values, beliefs, practices, etc., of "the Other," but thereby suggests that the members of one culture can never learn or gain anything of value (ethical or otherwise) from the members of another culture. But the history of how diverse cultures have emerged over time – i.e., precisely through processes of intermixing and hybridization with others – shows this to be false:

> If there were really an isolating barrier, of course, our own culture could never have been formed. It is no sealed box, but a fertile jungle of different influences – Greek, Jewish, Roman, Norse, Celtic and so forth, into which further influences are still pouring – American, Indian, Japanese, Jamaican, you name it. The moral isolationist's picture of separate, unmixable cultures is quite unreal. ... Except for the very smallest and most remote, all cultures are formed out of many streams. All have the problem of digesting and assimilating things which, at the start, they do not understand. All have the choice of learning something from this challenge, or, alternatively, of refusing to learn, and fighting it mindlessly instead. (Midgley [1981] 1996: 119)[1]

Especially as digital media dramatically accelerate these processes of encountering other cultures, we can indeed see rapid cultural change in our own day – described in part in terms of cultural hybridization and the development of "third cultures." (Such third cultures represent an amalgamation of elements from two or more diverse cultures that come into contact with one another. So we saw in chapter 2, for example, youth in Asian countries insisting on a more Western-like individual privacy, presumably under the influence of their exposure to Western cultures through diverse media, including the Internet and the Web. The resulting sense of privacy remains marked by indigenous roots in Confucian thought, Buddhism, Thai tradition, etc. – so that the resulting sense of privacy remains somewhat limited in comparison with Western conceptions. But this gives us a third culture of privacy, one that blends elements of two distinctive cultures into a unique combination that begins to take on a life of its own.) While digital media confront us with a seemingly overwhelming range of cultural diversity – thus dramatically heightening the temptation towards ethical relativism – the realities of contemporary cultural hybridization reinforce Midgley's argument. Insofar as ethical relativism leads to moral isolationism and a perhaps fatal paralysis of moral judgment,

---

1 For a more complete summary of Midgley's arguments, see <http://www.drury.edu/ess/values/MMidgley.html>.

these logical outcomes fly in the face of what we actually do in the contemporary world: we evaluate and make *judgments* about those elements of cultural practices, beliefs, norms, etc., different from our own that we will accept or reject.

Ethical absolutism (monism)

At the polar extreme of ethical relativism is a position often called ethical absolutism or ethical monism. Briefly, this view insists on the following:

> There *are* universally valid norms, beliefs, practices, etc. – i.e., such norms, beliefs, practices, etc., define what is right and good for all people at all times and in all places.

What is often tacit or unstated for the ethical absolutist is the additional claim:

> I/we *know* what those norms, beliefs, practices, etc., are – completely, clearly, unequivocally.

This may seem like an odd claim to spell out, but, as we will see, this is an especially crucial element of the ethical absolutist's position.

Finally, the ethical absolutist will thereby have to argue:

> Those norms, beliefs, practices, etc., that are *different* from the ones we *know* to be universally valid must therefore be wrong (evil, invalid, etc.).

In this way, the ethical absolutist is in the position to both applaud those beliefs and behaviors that agree with his or her own view of what is universally valid, and to condemn those beliefs and behaviors that differ from his or her own.

Given this meta-ethical framework, the ethical absolutist enjoys at least one advantage over the ethical relativist: the ethical absolutist *is* in a position to coherently applaud or condemn the values, beliefs, practices, etc., of others – e.g., s/he could applaud a Mother Teresa in India and condemn a Hitler in Germany. At the same time, however, this leads, obviously, to the *intolerance* of diversity that the ethical relativist finds so distasteful and destructive (and rightly, at least up to a point).

The contrasts between the ethical relativist and the ethical absolutist usually work around what we can call first-order ethical norms, values, practices, etc. – e.g., matters of abortion and euthanasia, war and peace, sexual identity/ies and relationships, freedom of expression, our treatment of animals and the environment at large, the role of the law vs. individual conscience, etc. For example, one could take an absolutist position either for or against abortion. An ethical absolutist might hold that all life is sacred – and that the baby/fetus in the mother's womb is a sacred life that must be protected at all costs, including, unfortunately, the cost of the life of the mother in certain circumstances. And hence, abortion is never justified, even to save the life of the mother. Another ethical absolutist might agree that all life is sacred – including that of the mother; and so, if, say, a monstrously deformed baby/fetus thereby directly threatens the life of the mother, it is morally permissible – indeed, morally required – to remove and destroy the baby/fetus for the sake of saving the mother's life. While the two absolutists will thus profoundly disagree with one another, an ethical relativist will say, in effect, to each his or her own; neither position is ultimately "right," but we should learn to tolerate important ethical differences such as these and go on.

Suffice it to say that the ethical relativist's response here will satisfy neither of our ethical absolutists. But for our purposes, the primary point at this juncture is to move to the second-order or *meta-ethical* level of discussion – i.e., to apply these meta-ethical positions to the ethical frameworks of utilitarianism and deontology. Hence we can ask: how would these two positions have us respond to the *differences* between *utilitarian* and *deontological* approaches? (As you might guess, we will ask the same question with regard to further differences as we explore additional ethical frameworks below.)

Roughly, it would appear that the ethical absolutist would require us to accept one of these approaches – and thereby reject the other. The ethical relativist, by contrast, would likely say: it doesn't matter – neither view can claim universal validity. Indeed, it's a waste of time to wrestle with this question, since there is no ultimate right or wrong in any event – it's all a matter of culture, individual preference, etc.

Reflection/discussion/writing: relativism and absolutism

1. Given the accounts of ethical relativism and ethical absolutism, which of these positions better describes your own with regard to the following (first-order) ethical claims and issues?

    A. The destruction of human life – and most especially innocent human life – is always wrong; hence, abortion is never justified.
    B. Our right to determine what happens to our own bodies is the most fundamental of human rights. Hence, a woman has an absolute right to determine what happens to her body – and this includes the right to abortion, especially if her own life is imperiled by a pregnancy.
    C. Killing is always wrong – even in self-defense.
    D. Killing is sometimes justified – beginning with self-defense.
    E. You should always keep a promise.
    F. Sex before marriage is morally acceptable.
    G. [suggest additional "hot-button" moral issues for discussion and reflection]

2. In response to these – and/or other – issues, it is probable that you will find that you are an ethical relativist with regard to some, and an ethical absolutist with regard to others.

    Insofar as this is the case, can you begin to sort out and articulate what *arguments, evidence,* and/or other sorts *reasons* you might have for supporting your position (i.e., as either an absolutist or a relativist) vis-à-vis a given issue?

Beyond relativism and absolutism: ethical pluralism

At this point, I hope it is beginning to be clear that, whatever the strengths and advantages of both ethical absolutism and ethical relativism, neither position is fully satisfactory. To begin with, if the previous reflection and writing exercise has been successful, it should have helped you discover that – like most people in my experience – there are ethical issues about which you may be profoundly absolutist, and others that seem to be best left to a sort of relativist tolerance (if not indifference).

But this is not especially coherent: ethical relativism and ethical absolutism make mutually exclusive claims – there are/are *not* universally valid norms, values, practices, etc. How can we coherently hold both of these claims together?

As you've likely guessed, there is a third position – *ethical pluralism* – that seeks to resolve some of the problems faced by relativism and absolutism.

At its most basic, ethical pluralism argues that the ethical absolutist *may* be right – at least with regard to his or her opening premise: there *are* values, norms, practices, etc., that are valid for all human beings at all time and all places. But the pluralist quickly parts company with the absolutist on a second point: rather than insisting that there is thus a *single* set of values, norms, practices, etc., that apply in exactly the same way at all times and all places, the pluralist argues that it is possible (indeed, inevitable and desirable) to *interpret/understand/apply* these norms in *diverse* ways in *diverse* contexts. In this way, the ethical pluralist is able to at least partially agree with the empirical observation highlighted by the ethical relativist. Obviously, we *do* observe that there *are* different practices in diverse times and cultures. But rather than interpreting these different practices as evidence for the absence of universally valid norms and values (as the relativist's argument does – and invalidly, as we have seen), the ethical pluralist argues that these diverse practices are the result of how different contexts will require us to interpret and apply the same norm in sometimes strikingly different ways.

To use a favorite example: it is easy to observe that people with kidney disease are treated differently in different cultures and places. In the United States – at least for those who can afford good health insurance – kidney dialysis, despite its enormous expense, will be made available more or less without regard for the patient's age. By contrast, the national health care system of the United Kingdom has set an upper age limit of 65 on patients for whom it will subsidize such treatments (Annis 2006: 310). Lastly, at least early in the twentieth century, in the harsh environment of the Canadian arctic, an elderly member of the Kabloona community who was no longer able to contribute to the well-being of the community might voluntarily commit a form of suicide (Boss 2005: 9f.; see Ess 2007b).

Again, the ethical relativist will argue that these three different practices clearly show that there are no values or norms shared universally across cultures. But for the ethical pluralist, these three practices stand as three diverse *interpretations, applications,* and/or *judgments* as to how to apply a single norm – namely, the health and well-being of the community – in three very different environments and cultures. So, at least the relatively affluent in the U.S. can afford the health insurance that will provide kidney dialysis without age limit; but even in a relatively wealthy nation such as the U.K., failure to set limits on subsidized treatments would quickly bankrupt the nationalized health system. Finally, in the unforgiving environments of the Kabloona, the well-being of the community would be jeopardized if scarce resources were diverted to caring for those who no longer could contribute to the community. Hence, such care is literally not affordable by the community, nor, apparently, expected by the individual. So, the *practices* of each of these communities clearly differ. But for the ethical pluralist, these different practices rest upon a basic *agreement* on the well-being of the community as a shared norm or value. Each practice, simply, represents a distinctive interpretation of that norm; the diverse contexts of these communities require each of them to interpret and apply that norm differently.

In this way, the ethical pluralist can agree with the ethical relativist that (a) we do observe diverse practices as we move through different cultures and times, and that (b) we should *tolerate* these differences – rather than condemn them straight out, as the ethical absolutist is forced to do – at least insofar as we can understand them to be different interpretations of a shared norm or value. But the ethical pluralist, unlike the ethical relativist, does not thereby tolerate any and all practices. (Recall: such tolerance entails for the ethical relativist a serious logical contradiction.) On the contrary, if a practice – e.g., genocide – can be shown to violate a basic norm or value (in this case, the well-being of the community, at least as understood as an inclusive *human* community, rather than an exclusive tribal community), then the ethical pluralist can condemn such a practice as immoral.

And so, the ethical pluralist can overcome some of the chief difficulties of ethical relativism, including its logical incoherence and its inability to distinguish between a Mother Teresa and a Hitler. At the same time, however, the ethical pluralist shies away from the sort of intolerance for difference that often follows from ethical absolutism. To recall: the ethical absolutist seems restricted to one and only one set of values and norms that must be interpreted, applied, and practiced the same way by all people in all places and all times – and so, any variation from this one set of norms and practices must be rejected as morally wrong. (In the example of kidney dialysis, a moral absolutist located, say, in the U.S., might then well condemn the practices of the Kabloona as immoral.) By contrast, the ethical pluralist can *tolerate* – indeed, endorse – these differences in practice, insofar as they can be shown to reflect diverse interpretations and applications of a shared norm or value.

In these ways, ethical pluralism seeks to take up at least a limited version of the tolerance for difference enjoined by the ethical relativist, while avoiding a tolerance so complete as to paralyze ethical judgment entirely. An ethical pluralist does so while at the same time taking up at least a limited affirmation of universally valid values, norms, practices as endorsed by an ethical absolutist, while avoiding the ethical *monism* and intolerance of difference that such absolutism easily falls into.

Strengths and limits of ethical pluralism. Ethical pluralism thus provides us with an important way of understanding and responding to the sometimes radical differences that we encounter, especially at a global level.

To put it negatively: if we can only choose between ethical relativism and ethical monism, then any effort to undertake a digital media ethics that might "work" cross-culturally would seem doomed to two equally unattractive choices: either we follow the relativist and tolerate any and all practices (saving us, it must be admitted, the difficult work of having to think about any of this at all . . .), or we adopt an absolutism that would result in a kind of ethical colonialism, i.e., the imposition of a single set of practices upon all peoples, because any *difference* from *the* right set of values and practices must be wrong.

To put it positively: ethical pluralism allows us to see – in some important cases, at least – how people in diverse cultures may share important norms and values; but at the same time, we are able to interpret and apply these norms and values in sometimes very different sorts of *practices* – ones that reflect our own cultural contexts and traditions. This means that ethical pluralism allows us to have a *global* digital media ethics – one that provides a shared set of guidelines for how we may ethically behave in relationship with one another. But these shared norms and values are interpreted through the lenses of different traditions and applied in different cultural contexts. These different interpretations or applications thereby allow us to preserve the practices and characteristics that make each culture distinctive and unique. In this way, ethical pluralism is a crucial element of the "ethical toolkit" we need if we are to develop a global ethics that respects and preserves diverse cultural traditions and identities.

Ethical pluralism enjoys two additional strengths. First, it is a way of approaching ethical matters that is found not only within Western traditions (beginning, at least, with Plato and Aristotle, but extending into contemporary ethical frameworks such as feminism [see Warren 1990]), but also in fact throughout diverse religious and philosophical traditions such as Islam (Eickelman 2003), Confucian thought (Chan 2003), and others. This is to say that ethical pluralism appears to be a widely shared and recognized way of approaching ethical differences – not simply a provincially Western way.

Second, ethical pluralism appears to in fact "work" in contemporary practices. Perhaps the most important example here is the issue of *privacy*. As we have seen in chapter 2, expectations of privacy and correlative data privacy protection laws vary from country to country – in part as they rest on dramatically different, if not contradictory, understandings of human beings. But it is arguable that there is increasing recognition of a shared notion of privacy that holds for both Western and non-Western countries and cultures. This shared notion is interpreted and applied in different ways – reflecting first of all the differences between cultures in terms of the importance they place on the individual vis-à-vis the community. The diverse *practices* of data privacy protection thereby reflect – and, more importantly, preserve – some of the fundamental values and traditions of each culture. In this way, ethical pluralism seems to "work" as an important component of a global information and computing ethics (ICE). And so we might expect that in other issues of digital media ethics, pluralism will likewise emerge as an important strategy for preserving cultural differences while developing a shared, genuinely global ethics.

At the same time, however, ethical pluralism will *not* resolve all the differences we encounter as different cultures and traditions approach the ethical issues of digital media. To use the example of the Muhammad cartoons (discussed in ch. 4), for at least many (though by no means all) religious believers, cartoons that can only be seen as blasphemy must *not* be published. For the editors of the Danish newpaper *Jyllands-Posten*, however, essential ethical and political values were at stake in commissioning and publishing the cartoons – namely, freedom of expression and freedom of the press. Add to this the cultural observation that for most Danes, *anything* – even the Queen – is an appropriate occasion for humor (at least, up to a point). It is by no means clear how the conflict here can be resolved in a pluralist fashion. Such an analysis would have to show that these two views are in fact *not* as contradictory as they appear – that they are rather simply diverse interpretations of a shared ethical norm (which one[s]?). (Additional critiques are offered by Hiruta 2006 and Capurro 2008.)

Hence, in the face of diverse cultural norms, beliefs, and practices, we will not always be able to resolve these sometimes deep and irreducible differences by way of an ethical pluralism. More broadly, then, in the face of such differences, we are obliged to discern whether we most justifiably understand and respond to these differences as an ethical relativist, an ethical absolutist, and/ or as an ethical pluralist.

Reflection/discussion/writing questions: meta-ethics – a first run

As many of the examples we've explored in this book should make clear, the *culture(s)* which surround us, whether in our upbringing and/or in our work and leisure as mature people, play a central role in shaping our ethical thinking. (At the same time, readers should keep in mind here the important caveats and difficulties of using cultural generalizations: see ch. 2, *Interlude*, pp. 39–43).

In particular, the comparative ethicist Bernd Carsten Stahl notes that since the twentieth century, at least within the English-speaking world, utilitarian approaches have dominated over alternatives. By contrast, deontological approaches – especially as rooted in Kant and then the contemporary German philosopher Jürgen Habermas – have been favored in the Germanic countries, including much of Scandinavia. These in turn contrast with what Stahl characterizes as French moralism in Montaigne and Ricoeur. On Stahl's analysis, this approach to ethics is *teleological*, i.e., oriented towards the goal or *telos* of discerning and doing what is necessary for the sake of an ethical and social order that makes both individual and community life more fulfilling, productive, etc., through "the propagation of peace and avoidance of violence" (Stahl 2004: 17).

As we will see more fully below, these views in turn contrast with non-Western traditions. Briefly (and somewhat misleadingly), modern Western traditions emphasize the *individual* as the primary agent of ethical reflection and action, especially as reinforced by Western notions of *individual* rights. Certainly, these traditions further recognize that individuals' actions are made within and affect a larger *community*; and, as we will see, there are ethical traditions in the modern West that indeed emphasize greater attention to *community*, not simply *individual*, action and good. But at least in comparison with modern Western traditions, non-Western traditions – including various forms of Buddhism, Confucian thought, and indigenous traditions in Africa, Australia, and the Americas – lay greater emphasis on the *community* and community well-being as the primary focus for ethical reflection and choice.

This ethical map becomes even more complicated, first of all, as we recognize that these generalizations will only go so far: again, each cultural generalization immediately implies counterexamples, additional layers and influences, etc. The complexity grows further as we add both (a) pre-modern and contemporary ethical traditions – as we are about to see, the *virtue ethics* expressed by Socrates and Aristotle and its contemporary expressions – and (b) contemporary ethical frameworks such as feminism and especially the *ethics of care*, along with *environmental ethics*.

While overwhelming at first, exploring these diverse ethical approaches is both (a) unavoidable, especially as digital media allow more and more people around the globe to communicate and interact with one another, and (b) necessary – first of all in order to overcome our own ethnocentrism and its attendant dangers. Such exploration should further help us make better informed choices regarding our own ethical frameworks and norms – and, ideally, help us move towards a more inclusive, genuinely *global* digital media ethics that recognizes and fosters our ethical differences alongside our shared norms and values.

At this stage, however, it may be helpful to pause to take a first run at learning how to apply the meta-theoretical positions of ethical relativism, monism, and pluralism.

I. Presuming your own prevailing cultural context(s) and/or culture(s) of origin are primarily Western, review Stahl's characterization of various national cultures as primarily utilitarian, deontological, and teleological.

A. Which, if any, of these frameworks seems *closest* to what you observe in your culture to be a prevailing way of making ethical decisions?

Illustrate your response with an example or two – ideally, one drawn from an ethical issue evoked by the use of digital media.

B. Which, if any, of these frameworks seems *furthest* away from what you observe in your culture to be a prevailing way of making ethical decisions?

You can illustrate and support your response here by applying this framework to the example(s) you describe in (1)(A).

C. What are the results? That is, do the two frameworks that you identify and apply in (1)(A) and (1)(B) issue in *conflicting* ethical conclusions (e.g., undertaking otherwise illegal music downloading because the benefits of doing so seem to outweigh the costs – i.e., a *utilitarian* analysis – vis-à-vis rejecting such an activity because it violates what may be argued to be a just law – i.e., a *deontological* analysis)?

And/or: do these two frameworks end up endorsing the same, or least coherent and complementary, ethical conclusions or claims? (For example, we saw in ch. 2 how both deontological and utilitarian approaches to privacy in the West endorse individual privacy rights as essential – though for characteristically different reasons.)

And/or: do these two frameworks issue in (at least, seemingly) *contradictory* results?

D. Especially if these two frameworks issue in different, perhaps contradictory, results, how do you respond?

That is: do you interpret or understand these differences primarily as

(i) an ethical relativist?
(ii) an ethical monist?
(iii) an ethical pluralist?

However you respond to these differences, do your best to support and justify your answer with one or more arguments, elements of evidence, etc.

2. The same set of questions – but now encompassing a global range of ethical frameworks – may be asked. In particular: if your cultural context(s) and/or culture(s) of origin are non-Western, so that you already have a strong familiarity with

especially non-Western ethical frameworks, now might be a good time to undertake the more global version of these questions. (And/or: you and/or your instructor may decide it's better to wait on these until further review of the discussion of these frameworks that is about to follow.)

Either way, this exercise should begin by asking you to take up two frameworks – one characteristically Western (e.g., utilitarianism) and one characteristically non-Western (e.g., Confucian, Buddhist, Hindu, African, etc.). With these two frameworks as your starting point, the questions in (1) can then be pursued.

4. Feminist ethics

As the discussion so far demonstrates, virtually all of the philosophers who have developed important ethical frameworks in Western (and, as we will see, Eastern) traditions are men. Especially for the second-wave feminists of the 1960s and 1970s, this observation naturally leads to an important question: is it possible that the conceptions, approaches, values, etc., that make up prevailing ethical (and other philosophical) frameworks *reflect* characteristically "male" or "masculinist" ways of knowing and thinking? Or, to state it negatively: is it possible that these prevailing ethical frameworks thus tend to ignore or exclude what are characteristically women's ways of knowing and reflecting on ethical issues?

In the domain of ethics – specifically, in the area of developmental psychology concerned with how people reflect on and seek to resolve ethical difficulties – these questions were given particular force through the work of Carol Gilligan. Gilligan's landmark book, *In a Different Voice* (1982), documented both important parallels and distinctive differences between the ways in which men and women characteristically approached important ethical dilemmas. Briefly put, Gilligan's interviews with women facing difficult ethical choices (including the possibility of abortion) challenged the then-prevailing schema of ethical development established through the work of Lawrence Kohlberg – work that, in fact, built on observations of and interviews with men exclusively. On the one hand, for both Gilligan and Kohlberg, the evidence of their inter-views and observations suggested that individuals develop their abilities to recognize and come to grips with ethical issues over time and in ways that can be described by a three-stage schema (with each stage in turn involving two sub-stages). *Pre-conventional* morality, describing how pre-adolescents grapple with ethical matters, works on a simple reward–punishment schema: one is "good" because good acts are rewarded, and one (usually) avoids being "bad" because bad acts are punished. *Conventional* morality, characteristically the moral stage of young adolescents and adults, reflects the values, practices, and expectations prevailing in the larger society, with an emphasis on *justice* and correlative notions of recognizing and preserving basic *individual rights* – at least as these contribute to the maintenance of the status quo. *Post-conventional* morality, by contrast, represents a move into significant sorts of ethical *autonomy* (in Kant's term), as individuals take conscious responsibility for their ethical principles and reflections in new ways, so as to perhaps radically critique and re-evaluate prevailing social claims regarding rights and justice. As is often the case, such reflections can lead individuals to draw new ethical conclusions regarding right and wrong that run *against* the prevailing morality of their larger society. Historically, such post-conventional moral-ists have been important for what we think of as ethical and social progress: their post-conventional morality has led them to chal-lenge prevailing social practices and values and, in the view of subsequent generations, helped lead society more broadly to a set of values and practices that are seen as ethically preferable over earlier ones. (To be sure, as the experience of these exemplary thinkers makes clear, moving to a post-conventional stage is difficult; indeed, Kohlberg claimed that most people never move beyond the conventional stage.)

While her findings support the outlines of this large framework, Gilligan found that as women moved through these stages, their moral experiences demonstrated important differences. For our purposes, the most important differences are as follows. For Kohlberg (and, to be fair, for most ethicists in the modern West), the key to moving beyond conventional morality is the critical use of *reason* – where reason is understood to focus especially on general principles, including rules of social justice and individual rights. So a Martin Luther King, Jr., for example, can argue that segregation laws are unjust because they violate the basic principle of justice in a democracy and the modern liberal state; only those laws are just that rest on the consent of the governed. But segregation laws were passed by a white population, in states where the people of color also affected by these laws had no vote – and hence no possibility of exercising consent. Hence such laws are unjust. On the basis of such arguments, King can then justify *disobeying* the law of the land – in developmental terms, going beyond conventional morality to a post-conventional morality based on clear principles of justice and rights (King [1963] 1964).

To be sure, Gilligan found that women certainly employ reason – minimally, the capacity for inference and the recognition of important general principles – in confronting their ethical quandaries. But in addition to reflection on general principles, Gilligan found that women as a group tended to make three distinctive maneuvers. To begin with, as Piaget had already observed, little girls may be less concerned than their male counterparts with making sure, for example, that all the rules of a game are followed (*justice*), while they be more concerned that everyone within a given group has the *feeling* of being treated fairly, of being included, etc., even if this sometimes means breaking the rules (Gilligan 1982: 32–8). But this means, second, that women as a group tend to focus on the *emotive* dimensions of an ethical problem. Third, a problem is seen to be ethical especially as it involves a web of interpersonal *relationships*, not simply individuals as "nodes" in those relationships marked only by defined sets of rights, etc.

So, for example, Kohlberg asked his (male) interviewees to respond to the "Heinz dilemma." In this scenario, a husband (Heinz) needs to get life-saving medicine for his wife; but he cannot afford to do so, and so his pharmacist refuses to provide him with the medicine. In Kohlberg's analysis, men as a group tended to analyze this dilemma in terms of the rights and principles involved – e.g., the right of the pharmacist to protect his property (and sources of profit and livelihood) vs. the wife's ostensible right to life. But as young women were presented with this dilemma, as a group they tended to want more information – first of all, about the *relationships* between the three protagonists. For example: would Heinz's wife really want him to risk going to jail for her sake? Is it possible that they could talk with the pharmacist and work out a way to pay for the drug over time (Gilligan 1982: 25–32)?

In these ways, the women's questions often teased out specific details about the possibilities and relationships in play that might otherwise be ignored through an exclusive focus on general principles of justice and abstract rights. In doing so, the women's questions may suggest alternatives to the simple, "either/or" dilemma presented at the outset – i.e., either respect the law (and lose your wife) or disobey the law (but save your wife). So, as some of my own students have suggested: if the pharmacist is a friend who knows and trusts Heinz and his wife, why couldn't he arrange for Heinz to pay for the needed drug over time, rather than insisting on an all-or-nothing payment?

For Gilligan, women's ethical development could thus be characterized as an ethics of *care* and responsibility for both others and oneself (the latter, at least, in the post-conventional stage) – in contrast with (but not in opposition to) the ethics of principles, rules, and justice that characterized the ethical focus of many (but by no means all) men. Finally, Gilligan emphasized that these two patterns of ethical development, while clearly different, are not mutually exclusive. Rather, both patterns are essential – and, ideally, conjoined in a synthesis that holds both together.

Of course, there are any number of controversial and highly contested assumptions and claims at work here, as the subsequent development and debates regarding feminist ethics bring to the forefront. To stick with just one: does Gilligan's schema run the risk of *essentialism* – of assuming or arguing that there is something (an "essence") about being a biological female that strongly directs (or simply determines) that all women must follow the lines of ethical development articulated in Gilligan's schema? Because such essentialism risks throwing us back to gender stereotypes that have been used throughout the history of patriarchy to justify women's subordination to men, it is argued to be a mistaken assumption from the outset. And Gilligan would deny that she is making such an essentialist assumption. But it remains a very great problem for feminist theorists both to avoid such essentialism and simultaneously to seek to argue and claim that women as a group tend to think, feel, etc., in ways that are distinct and legitimate alternatives to the ways more characteristic of their male counterparts.

Despite these and related difficulties, however, Gilligan's work inaugurated important new developments in ethical theory, beginning with greater respect for the positive role of emotions – specifically, *care* – as developed more extensively, to begin with, by Sarah Ruddick (1989) explicitly in terms of an *ethics of care*. To be sure, one does not have to be a feminist to take up an ethics of care: early on in the modern West, David Hume famously argued that ethical reflection is fully reducible to emotions; but for some of us, this goes too far, especially as it runs the risk of thereby reducing all ethical claims to purely *relative* ones.

Despite this risk, as we will see again in the context of *virtue ethics* (section 5, below), there is a growing recognition from a variety of sources – feminist ethics, virtue ethics, neurobiology, and comparative philosophy more broadly – of the central roles emotions play in ethical decision-making. For example, some evidence suggests that a lack of emotion, as resulting from certain forms of brain damage, makes it impossible for people to *make* ethical decisions in their daily lives. Simply put, our internal sensibilities regarding the relative ethical importance of various possible acts (i.e., how far an act or choice may be good, really good, a little evil, really evil, etc.) appear to be *felt* as much as they may be rationally sorted out in a kind of calculus. So if our capacities for emotion are damaged, we are no longer able to sort through the ethical choices confronting us: "Without an affective element the agent will be unable to rank the items to be judged in order of their significance to her directly, or to her indirectly by the affect they are likely to have on those about whom she is concerned" (Stuart 2007: 144). What is striking about these turns towards recognizing the integral role emotions play in our decision-making process is that these insights thereby point us both towards pre-modern Western understandings of our ethical life as involving *both* thought and feeling (e.g., in the Socratic and Aristotelian conception of *phronesis,* a practical ethical judgment that is *felt* as much as thought) as well as towards non-Western understandings – e.g., the Confucian view of the human being as incorporating *xin,* what Ames and Rosemont translate as "heart-and-mind," to make the point that "there are no altogether *disembodied* thoughts for Confucius, nor any raw feelings altogether lacking (what in English would be called) 'cognitive content'" (1998: 56 – emphasis added). The role of emotions in ethics is thus a shared understanding across a literally global scale; as feminist ethics brings this role to the foreground, it thereby points towards what may be a "bridge" concept, a shared understanding between both Western and Eastern views that will play an important role in any global digital media ethics.

Moreover, in emphasizing the importance of webs of interdependent relationships, in contrast with prevailing emphasis on *individual* rights, feminist ethics thereby supported and developed alongside (then) new forms of *environmental* or *ecological* ethics. Briefly, such ethics extends the modern Western focus on the rational individual human being as the primary *moral agent* who deserves *moral status,* so as to argue that non-human entities, including not only living beings but the larger ecological systems they constitute in relationship with the natural order, also deserve and require moral status and respect in our ethical reflections.

In these ways, feminist ethics helps us move to a more inclusive and comprehensive account of *how* we may come to grips with the ethical challenges we face.

Applications to digital media ethics

Arguably, an ethics of care is at already at work in a number of choices and behaviors associated with digital media. For example, using one's mobile phone to simply check in on friends and loved ones – "just to say 'hi'" – seems like a clear expression of care and concern, one that nicely reinforces existing relationships.

As we've also seen, for those who enjoy using digital media to copy and distribute songs, videos, etc., that they enjoy, "sharing is caring." That is, it would appear that a primary motive in such sharing is our pleasure in giving to friends and loved ones the chance to enjoy the same music and videos that we have enjoyed.

These examples, however, also highlight one of the important limitations to an ethics of care. Insofar as such an ethics stresses the importance of our emotional bonds with one another, it thereby runs the risk of restricting our ethical focus too narrowly, i.e., upon a relatively small circle of family, friends, and loved ones. Taken to its extreme, a care ethics could thus justify our ignoring whole populations around the globe, because, simply, we do not experience a relationship of care with such populations. But in a world ever more interwoven via digital media – unless these media help us learn how to *care* for others beyond our immediate circles – the ethic of care runs the risk of an increasingly inappropriate provincialism.

Reflection/discussion/writing questions: feminist ethics and digital media

1. Choose an important ethical dilemma concerning the use of digital media – either one that has been explicitly articulated and addressed in this book, or one of your own choosing.

    [So, for example, you may recall the scenario from the beginning of chapter 3 – one that raises the possibility of stealing a desired CD from a music store.

    The dilemma presented there appears to be a simple either/or: either steal the CD and enjoy the music (but risk possible legal problems, including fines and jail time), or don't steal the CD and fail to enjoy the music (but avoid possible legal problems, including fines and jail time).]

    A. Identify the dilemma carefully, making sure that you identify as much of the ethically relevant values, principles, etc. that you see coming into play.

    B. After reviewing the account given above regarding possible contrasts between an ethics of rights/principles and an ethics of care and responsibility, does your own analysis of the dilemma you have identified tend to follow one or the other (possibly both) of these ethical approaches?

C. In particular, in your analysis, do you tend to remain with an "either/or" conflict that seems to be generated by considerations of the *principles, rights,* etc. that come into play here; and/or are you further inclined to want to know more specific details regarding the context of a given ethical choice – including, for example, the sorts of relationships and feelings at work?

D. Insofar as these sorts of differences emerge – perhaps in a group and/or class discussion – do they seem to be correlated with gender in the ways Gilligan has claimed? If so, what might this correlation suggest for how we think about "doing ethics"? That is, *are* there important differences between masculine and feminine approaches to ethics – and if so, what is our *ethical* responsibility for taking these differences into consideration?

2. In my view, one of the most important contributions of feminist ethics and an ethics of care is not only that they require us to acknowledge the significance of emotions, including feelings of care, but also that they help us learn to think beyond more dualistic, "either/or" approaches that have been emphasized in modern Western reflection and teaching about ethics. By moving towards a "both/and" logic (or logic of complementarity), in particular, we are sometimes able to see third (or more) alternatives and possibilities – overlooked by more dualistic ways of thinking – that thereby may help us resolve what otherwise seem to be intractable dilemmas of the sort faced by Heinz.

These (for the modern West, new) ways of thinking, moreover, are valuable not only as they help sustain a much needed environmental ethics, but, further, as such relational thinking may closely resonate with (i) contemporary non-Western ethical frameworks (explored more fully below) and (ii) especially the *networked* or *distributed* character of ICTs and other digital media linked together through the Internet and the Web.

A. Given what you are able to understand about these two different *logics* – a logic of dualism as based on the exclusive "either/or" and a logic of complementarity, of "both/and" (discussed in ch. 5, pp. 139–41), as you observe the larger culture around you, which of these two logics appears to be at work more predominantly than the other? Be sure to provide an example or two to help illustrate your point.

B. Identify a central issue in digital media ethics that you have already analyzed and responded to with some care in the course of your working through this volume. Review your response: do you seem to rely on one of these logics more than the other in your analyses and resolution(s) of this issues? Be sure to explain carefully how the logic you identify is apparent in your analysis/resolution.

C. After reviewing your analyses and resolution(s), insofar as they seem to rest on using one logic more than another, would the analyses and resolution(s) be any different in any significant ways if you were to attempt making these analyses and resolution(s) using the other logic instead? If so, how? Be sure to explain carefully how this is so.

5. Virtue ethics

Virtue ethics is both ancient in the West (associated with especially Socrates and Aristotle) and global, in the sense that we find versions of virtue ethics in diverse philosophical and religious traditions around the world (including, as we will see in the next section, in Confucian and Buddhist thought). In this way, virtue ethics is an important common ground for ethicists from diverse traditions, one that has clear potential to serve as a significant component of a shared global ethics. Indeed, virtue ethics has enjoyed something of a renaissance in recent decades among Western philosophers for a number of important reasons – including precisely its potential for providing a common ethical ground for global ethics.

Virtue ethics begins with the sensibility that what we *ought* to do as human beings is, first of all, to become *excellent* human beings. *Becoming* an excellent human being, more precisely, means to develop and fulfill our most important capacities *as* human beings. Clearly, as individuals we may have a distinctive set of potential abilities, such as athletic or musical abilities. But for Socrates and

Aristotle, our most important abilities *as human beings* as such, not simply as individuals, are our capacities to *reason* – and this in two ways. What Aristotle (and later Kant) identified as the "theoretical" function of reason centers on what we now think of as a scientific understanding of the laws and principles that guide the workings of the physical world. For the ancient and medieval thinkers in the West, this capacity to understand reality was important for a number of reasons. In particular, by understanding reality properly, we as human beings can then "attune" ourselves to that reality: that is, we can better know what to expect of it, and better know how to behave within and in relationship with it, in order to achieve what the Greeks called *eudaimonia* – often translated as "happiness" but better understood as a kind of fundamental sense of well-being and contentment.

But if our goal as human beings is to achieve such contentment or *eudaimonia*, then it is equally important that we develop what Aristotle (and, subsequently, Kant) identified as *practical reason*. Such practical reason involves first of all our ability – given our best knowledge of reality and thus of our possible choices and actions – to make the sorts of analyses and ethical judgments required for us to do "the right thing" both for ourselves as individuals (the ethical for Aristotle) and for our larger communities (for Aristotle, the political). As we have seen, these sorts of ethical decision-making further require what Socrates and Aristotle term *phronesis* – a practical judgment that is able to discern the right choice (or, sometimes, choices) among the possibilities before us.

This capacity for judgment, we can notice, is one that is capable of learning from its mistakes. So Socrates (as related by Plato) uses the ship's pilot and the physician in *The Republic* as primary exemplars of people who exercise such judgment, and notes:

> . . . a first-rate pilot [*cybernetes*] or physician for example, *feels* [*diaisthanetai*] the difference between the impossibilities and possibilities in his art and attempts the one and lets the others go; and then, too, if he does happen to trip, he is equal to correcting his error. (*The Republic*, 360e–361a [Plato 1991] – emphasis added; cf. *The Republic* I, 332c–e; VI, 489c; X, 618b–619a/301)[2]

---

2 Following standard practice among Plato scholars, page references are to the Stephanus volume and page number.

And learning from mistakes means, as Aristotle emphasized, that our developing these capacities of ethical judgment and analysis, and of reason more broadly, is an on-going task: just as the athlete or physician must constantly practice if s/he is to maintain, much less improve, his or her abilities, so we as human beings must likewise cultivate in a conscious and on-going way our rational abilities, including our use of *phronesis*.

To put it somewhat differently: *being* a human being is not something that is simply given or taken for granted. Rather, *becoming* a human being – meaning, a being capable of (among other things) making the ethical and political judgments required for living a good ("happy") life in a community thereby marked by harmony and well-being – is an on-going task.

Finally, it is important to emphasize: while developing our other capacities – e.g., as athletes, musicians, lovers, friends, parents, game-players, etc. – is important, for Socrates and Aristotle it is very clear that there is nothing more important than the task of cultivating and practicing excellence *as a human being*, meaning, as a human being engaged with making ethical and political judgments and choices. In particular, if we subordinate our cultivation of excellence as ethical and political beings to any other activity – e.g., the pursuit of wealth or power – we thereby put our capacity for reason and ethical judgment at risk. Indeed, Socrates and Aristotle argue that if we allow our interests in wealth and power to persuade us to judge and act *against* our reason and better judgment, we thereby *harm* these capacities (just as we would harm a race horse, to use Socrates' analogy, by using it as a plow-horse instead). But if we harm and thereby diminish these capacities, we thereby undermine the capacities most central to our discerning what is genuinely good, pursuing it, and thereby achieving *eudaimonia* or well-being.

This is *not* to say, as some later moralists argued, that we can achieve *eudaimonia* only by abstaining from the pursuit of, say, wealth and power. Rather, Socrates and Aristotle are optimistic that both *eudaimonia*, as resulting from pursuing our excellence as ethical and political beings, and (at least a moderate amount of) wealth and power can be had together. (Indeed, for Aristotle, a moderate amount of wealth and power are necessary conditions of cultivating theoretical and practical reason, and thereby of achieving *eudaimonia*.) But the constant danger is to let our interests in wealth and power overshadow our pursuit of excellence as ethical and political beings – and thereby, to paraphrase Jesus four centuries later, to gain the whole world but lose our souls.

So Socrates (again as related by Plato) says in *The Apology*:

> It is God's bidding, you must understand that; and I myself believe no greater blessing has ever come to you or to your city than this service of mine to God. I have gone about doing one thing and one thing only, – exhorting all of you, young and old, not to care for your bodies or for money *above* or *beyond* your souls and their welfare, telling you that virtue does *not* come from wealth, but wealth from virtue, even as all other goods, public or private, that man can need. (*The Apology*, 29e–30b [Plato 1892] – emphasis added)

In this way, Socrates argues for the absolute priority of human excellence over all other interests if we are to achieve *eudaimonia* or well-being, but insisting thereby that our pursuit of excellence will also lead to the other human goods that we desire and need.

While deontology and consequentialism dominated much of the ethical discussion among Western philosophers in the twentieth century, within the last three decades virtue ethics has enjoyed a considerable revival. Rosalind Hursthouse nicely summarizes why: for all of their strengths, neither deontology nor consequentialism seems to address a number of topics required for a complete moral philosophy, including "moral wisdom or discernment, friendship and family relationships, a deep concept of happiness, the role of the emotions in our moral life, and the questions of what sort of person I should be . . ." (1999: 3).

All of these elements are important, in my view; but I would highlight here that an initial strength of virtue ethics is precisely the attention it gives to the *emotions*. As feminist philosophers made abundantly clear in their various critiques of received Western traditions, especially the modern emphasis on *reason* in ethical reflection rested on a terribly destructive dualism inherited from René Descartes ([1637] 1972, [1641] 1972), one that set mind radically apart from body – and thereby from emotions. This dualism both derived from and reinforced earlier gender stereotypes – roughly, men think, women feel – that not only worked to perpetuate the subordination of women in modern societies (and their exclusion from the academy, most especially philosophy departments), but also resulted in understandings of ethical reflection that systematically ignored and denigrated the emotional. Yet it seems that we often make good and justifiable ethical decisions in part as we pay attention to our feelings – "what my gut tells me," in idiomatic American. Indeed, it seems increasingly clear that we must pay attention to emotions and feelings. Briefly, emotions are part and parcel of what it means to help or harm a human being (i.e., such help usually results in some sort of pleasure, while harm usually involves some sort of pain). Moreover, emotions appear to play a critical role in *motivating* us to behave ethically in the first place (Gazzaniga 2005: 167). In addition, as we saw in the discussion of emotions in the previous section on feminist ethics, there is interesting evidence to suggest that a lack of emotion – e.g., as caused by specific forms of brain damage – incapacitates moral judgment. Emotions play a crucial role in our sense of what is important and what is not – i.e., we act on the basis of a *felt* set of values, not simply an intellectual calculus of comparative weights (see Stuart 2007). Finally, as we saw, this attention to emotion helps to bridge Western ethics with non-Western views, such as Confucian thought. Again, in contrast with the Cartesian mind–body split, Ames and Rosemont translate *xin* as "heart-and-mind," in order to emphasize that thought and feeling always accompany one another (1998: 56). As in the case of feminist ethics, when virtue ethics brings to the foreground the importance of emotions in our ethical lives, it thereby points to a post-Cartesian view – one that brings Western ethics closer to at least some of its non-Western counterparts. Doing so may be an essential step in the development of a more global digital media ethics – i.e., one that "works" in both Western and non-Western cultures and traditions.

Moreover, virtue ethics, as including a focus on the development of moral *judgment*, thereby highlights a critical element of learning how to be human – both alone and with others: most importantly, as it is only through developing and exercising such judgment that we can claim to be autonomous and (self-)responsible human beings. Without such judgment, simply, we are likely to only follow the dictates of others.

Finally, we have seen that some modern Western ethical frameworks starkly contrast with their non-Western counterparts. Aristotle's virtue ethics, however, resonates with similar emphases on *becoming* an excellent or exemplary human being as a focus of one's life as a *human being* that are found in a number of philosophical and religious traditions around the world, including Buddhism and Confucian thought. We will explore this more fully below, but here it is worth pointing out this resonance precisely in terms of the notions of *judgment* and self-correction: so Theptawee Chokvasin, for example, highlights in Buddhist thought the importance of *Attasammapanidhi*, "the characteristic of a person who can set herself in the right course, right direction in self-guidance, perfect self-adjustment" (2007: 78). For Chokvasin, this notion of self-guidance closely parallels modern Western notions of moral autonomy, specifically as articulated by Kant and Habermas. In addition to this important resonance with modern Western ethics, as Soraj Hongladarom and I have discussed, the Buddhist concept of *Attasammapanidhi* further closely resonates with the emphasis in Western virtue ethics, beginning with Socrates, on the central importance of moral judgment, especially as it allows us to take up the responsibility of choosing and guiding our actions as we navigate among the various ethical choices we face. (Socrates, in fact, expresses this point using the *cybernetes* – the pilot or steersman – as his model: cf. Hongladarom and Ess 2007: xix, xxiii, xxixf.)

Virtue ethics: sample applications to digital media

An initial way of applying a virtue ethics to digital media, as noted in the previous chapter, is to ask the question: what sort of person do I want/need to *become* to be *content* – not simply in the immediate present, but across the course of my entire (I hope, long) life? Along these lines: what sorts of *habits* should I cultivate in my behaviors that will lead to fostering my reason (both theoretical and practical) and thereby lead to greater harmony in myself and with others, including the larger natural (and, for religious folk, supernatural) orders?

As part of its resurgence in the contemporary West, virtue ethics has found wide application – including to such seemingly exotic topics as designing ethics for robots (e.g. Coleman 2001). In this volume, I have applied virtue ethics especially to the topic of cross-cultural communication online (ch. 4) and in a reflection/writing exercise regarding computer games in chapter 5 (see pp. 157–8).

Reflection/discussion/writing questions: the virtues of games

1. If you have not already responded to the reflection/discussion/ writing exercise in chapter 5 (see *A virtue ethics critique*, pp. 158–60), please do so now.

2. If you *have* already responded to that exercise:

    A. Review your responses to those questions.
    B. Would any of your responses change in light of this more extensive exploration of virtue ethics?
       ·If so,

       (i)   identify the specific questions/ideas/claims, etc., where you now respond differently than before, and
       (ii) explain as fully as you can *why* you now have a different response?

       In particular, are there new claims, arguments, insights that you did not have before that now lead you to draw different ethical conclusions? If so, what are these?

       If not, why not? That is, if your responses are the same as before, then

       (i)   identify one or more elements in this reading that are new to you, and
       (ii) why they do *not* lead you to change your mind.

       In particular, do these new elements force you to develop new arguments in order to sustain your already stated views, claims, etc.? If so, articulate one or more of these as fully as you can.

6. Confucian ethics

Confucian thought begins with a very different understanding of the human being than is held in modern Western theories. Modern Western thought strongly tends to assume that human beings are "atomic" individuals – i.e., that the human being as an individual is the most basic element or component of society, one that begins and can remain in complete solitude from others. (This atomism is traceable to the English philosopher Thomas Hobbes and the French philosopher René Descartes, but that story is too long to develop here.) Henry Rosemont (2006) has characterized this view as the "peach-pit" view of human beings. That is, a peach presents us with a surface – one that grows, changes, and finally dies over time. But underneath these surface changes remains the peach-pit – a stony, hard core that remains (relatively) unchanged over time. The peach-pit is thus closely analogous to traditional Christian and Islamic conceptions of the soul, and modern conceptions of the atomistic self. That is, underlying a surface body that grows, changes, and ultimately dies with time there is thought to be the "real" self, the *identity* that remains the same through time, "underneath" the outward and surface appearances of the mortal body. To be sure, this conception of the self resolves some important philosophical and ethical problems concerning identity – e.g., if there is no substantive, real self underneath the constant changes of a body, then who or what is *responsible* for that body's actions? That is, if the body associated with "you" committed a terrible crime five years ago, is it reasonable to say something like "that wasn't really *me* – I [meaning, my body] have *changed* and can no longer be held responsible for what I [my body] did five years ago"? Generally, in the modern West we *do* think that individuals remain responsible for their acts through time; thinking this way makes sense on the assumption of a "peach-pit" or atomistic self/identity that remains more or less the same over the life-course.

Such a conception of the self, however, can be understood as the result of a long development in Western societies, primarily in the last five hundred years, beginning with the Protestant Reformation. The Protestant emphasis on the *individual* soul and salvation is then philosophically refined and secularized in figures such as Descartes. Making *real* such a conception of the self further appears to depend on the wealth generated through industrialization. (As we have seen in the discussion of privacy, such a conception of the self, while initially alien to Eastern societies such as China, Japan, and Thailand, is becoming increasingly apparent – in part, as these societies develop the wealth that make individual privacy realizable, e.g., through the luxury of private rooms for children, etc.)

By contrast, in classical Confucian thought (and elsewhere, as we will see), human beings are understood first of all as *relational* beings: we *are* who we *are* always and only as we are taken up in specific relationships with others. For me, this means that I am *always* and *only* someone's son, brother, spouse, father, uncle, friend, employee, boss, beneficiary, etc.; and *how* I am – i.e., my choices, attitudes, behaviors, etc. – is always shaped in specific ways by each specific relationship. And so, how I am in relationship with my parents is different from how I am in relationship with my spouse, my siblings, my own children, my students, etc. To continue with Henry Rosemont's (2006) organic metaphors, in classical Chinese thought, human beings are like onions, not peaches: each of our distinctive relationships with others – including the larger social and political communities and, finally, the natural order at large (*Tian*) – constitutes one of the multiple layers that in turn make up who we are as human beings. In contrast with the peach-pit model, however, if we remove the layers of relationship from the onion, there's nothing left.

In ways closely analogous to the virtue ethics in the West, this understanding of the human being as a relational being means that ethics is primarily about becoming a (more) complete *human being* – first of all, by cultivating the behaviors and attitudes required for establishing *harmony* both among members of the human community (beginning with the family) and with the larger order (*Tian*) as such. In classical Confucian thought, this begins with learning and practicing *filial piety*, respect and care for one's parents, and ritual propriety. But the ultimate aim is to become an exemplary person (*junzi*) – someone who has cultivated and practiced appropriate attention to and care for others to such a degree that this exemplary behavior *is* who that person is. So Confucius describes the exemplary person as follows:

> The Master said, "Having a sense of appropriate conduct (*yi*) as one's basic disposition (*zhi*), developing it in observing ritual propriety (*li*), expressing it with modesty, and consummating it in making good on one's word (*xin*): this then is an exemplary person (*junzi*)." (15.18: Ames and Rosemont 1998: 188)

The exemplary person, in short, is one who has shaped his or her basic character or disposition through the practice of appropriate conduct and ritual propriety. The primary markers of such a character are modesty and integrity.

Much as Socrates and Aristotle emphasized achieving human excellence through cultivating and practicing the right habits throughout one's lifetime, Confucian ethics emphasizes that the project of becoming an exemplary person (always in *relationship* with others) is a life-long project. As one of the most famous of the Analects has it,

> At fifteen my heart-and-mind was set on learning.
> At thirty my character had been formed.
> At forty I had no more perplexities.
> At fifty I realized the propensities of *tian* (*T'ian-ming*).
> At sixty I was at ease with whatever I heard.
> At seventy I could give my heart-and-mind free rein without overstepping the boundaries.
> (2:4: Ames and Rosemont 1998: 76f.)

This is to say, for Confucius, cultivating the virtues or excellences, beginning with filial piety, leads to a sense of harmony or resonant relationship with both other human beings and the larger order or things – a sort of freedom and contentment that can be achieved in no other way.

And because, finally, it is believed that such ultimate freedom and contentment can be achieved only through the cultivation of excellence as a human being, we are always mistaken when we believe we will achieve happiness through other means instead, such as wealth and honor. So, just as Socrates and Aristotle later emphasized the importance of putting such human excellence first, in the same way Confucius insists that such excellence or virtue – for Confucius, following the proper *dao* or path – must always come first:

> The Master said, "Wealth and honor are what people want, but if they are the consequence of deviating from the way (*dao*), I would have no part of them. Poverty and disgrace are what people deplore, but if they are the consequence of staying on the way, I would not avoid them." (4.5: Ames and Rosemont 1998: 90; see also 4.11)

Confucian ethics and digital media: example applications

We have seen that Confucian ethics is at the center of a major conflict between Western and Eastern attitudes and practices regarding copyright (ch. 3). As a reminder, within a Confucian framework, an exemplary person, as benevolent towards others, would want to share the important insights that have allowed him or her to become such a person with others who likewise seek such excellence. Hence, the text he or she produces to record such insights is seen not primarily as a matter of personal property, but rather as a gift to be given to others – one that, indeed, may work as a kind of essential toolkit for the larger life-project of becoming an exemplary person. The appropriate response of those benefiting from this gift might include copying it and giving it to others – first of all, as a mark of respect and gratitude for the work of the exemplary person. In this light, copying and distributing a text is not primarily a matter of violating one's personal property as articulated in terms of copyright limitations; it is rather a matter of showing respect and gratitude for the gift of a benevolent master.

7. Additional perspectives – African

Colleagues engaged in the global dialogues on information and computing ethics represent a number of important linguistic/cultural domains – certainly Western perspectives (the U.S., U.K.,

Australia, northern and southern Europe, Scandinavia), as well as Asian perspectives (including China, Japan, Thailand, India). To my knowledge, there has been comparatively less representation and participation (at least in the English-language literature) from Latin American countries. For their part, however, African thinkers have recently become more engaged in these global dialogues – sparked in part by the first African Information Ethics conference, held in Pretoria, South Africa, in February 2007.

In his opening address to the conference, Rafael Cappuro emphasized the importance of *ubuntu* as an indigenous philosophical tradition and framework for developing an information ethics appropriate to the African context (2007: 6). As we saw in an introductory way in the discussion of Open Source and FLOSS (ch. 3), *ubuntu* (as inspiring the popular Ubuntu distribution of Linux) emphasizes that we are human beings in and through our relationships with other human beings: " . . . to be human is to affirm one's humanity by recognizing the humanity of others and, on that basis, establish humane respectful relations with them" (Ramose 2002: 644, cited in Capurro 2007: 6). While not all peoples and traditions in Africa recognize the term *ubuntu*, this notion of being human as involving an intrinsic interrelationship with and interdependence upon others is widely characteristic of African thought. So Barbara Paterson has observed that "[i]n African philosophy, *a person is defined through his or her relationships with other persons,* not through an isolated quality such as rationality [Menkiti 1979; Shutte 1993]" (2007: 157, emphasis added). And just as Confucian thought, in beginning with the person as a relational being, thereby stresses interaction with the larger community (both human and natural), so, Paterson continues, in African thought, in community, "[t]hrough being affirmed by others and through the desire to help and support others, the individual grows, personhood is developed, and personal freedom comes into being" (2007: 158). This means that *personhood* is not a given, but rather an on-going project: "African thought sees a person as *a being under construction whose character changes as the relations to other persons change.* To grow older means to become more of a person and more worthy of respect" (Paterson 2007: 158 – emphasis added). Again, given this concept of the individual, engagement with the community is paramount: "The individual belongs to the group and is linked to members of the group through interaction; conversation and dialogue are both purpose and activity of the community" (Paterson 2007: 158).

It would appear, then, that African traditions closely parallel both Confucian thought and Aristotelian virtue ethics, beginning with their shared emphasis on the individual human being as first of all engaged with the larger human (and natural) communities for the sake of both individual and community harmony and flourishing.

Hence, Confucian and Aristotelian approaches may provide helpful analogues for African thinkers as they explore and develop their own forms of information and computing ethics. But this exploration and development are just now emerging, and it will be very interesting to see where these developments take African philosophers and users of digital media – both for their own sake, and for the sake of the larger global dialogue regarding ICE and digital ethics more generally.

Reflection/discussion/writing questions: ethics and meta-ethics

Now that you have reviewed a global range of ethical frameworks – review one or two of the specific issues/cases of digital media ethics that you have analyzed and perhaps resolved with some care in the course of working through this volume.

1. Which of the ethical frameworks that we have now explored, i.e.,

    *   utilitarianism
    *   deontology
    *   feminist ethics/ethics of care
    *   virtue ethics
    *   Confucian ethics
    *   African ethics

    seem(s) to have been most in play in your reflections and decision-making?

Explain your response here with some care, making clear for yourself (and your reader, if applicable) *how* your analyses and resolutions fit the patterns and approaches of a given ethical framework.

2. Choose a framework that seems very far away from your own ethical starting points (identified in (1)).

    Take up this same ethical issue and, as best you can, provide an analysis and resolution of the issue using this alternative framework.

    How far are the results similar to and/or different from the results using your original ethical theory/ies?

3. How do you respond to these differences? That is, given what we've now learned about

    -   ethical relativism
    -   ethical monism/absolutism
    -   ethical pluralism

    which of these three meta-ethical frameworks are you most likely/able to apply to any *differences* that may emerge between the analyses and resolutions you have developed in (1) and (2)?

For further reading, research

Ethical theories

There are an enormous number of excellent textbooks that provide a more extensive introduction to and analysis of the ethical frameworks discussed here. If you are interested in learning more about these sorts of theories, the following will be useful:

Boss, Judith A. 2005. *Analyzing Moral Issues* (3rd ed.). Boston: McGraw-Hill. [Boss's text is exemplary for its introduction of various issues and approaches in ethics, with an especially strong emphasis on their application to current, real-world issues.]
Hinman, Lawrence M. 2008. *Ethics: A Pluralistic Approach to Moral Theory* (4th ed.). Belmont, CA: Wadsworth.

[In addition to providing excellent overviews of basic *theoretical* frameworks such as utilitarianism (Hinman's ch. 5), *deontology* (ch. 6), and *virtue ethics* (ch. 9), Hinman's text is useful for pursuing the *meta-theoretical* issues raised in this chapter regarding relativism, absolutism, and pluralism (his chs. 2 and 9; in the latter, he helpfully connects pluralism with Aristotle's virtue ethics and notion of *phronesis* – see esp. 285–8). Moreover, Hinman's "Ethics Updates" – <http://ethics.sandiego.edu/> – is one of the most comprehensive and genuinely useful online resources for both theoretical and applied ethics.]

Hursthouse, Rosalind. 1999. *On Virtue Ethics.* Oxford: Oxford University Press.
[Hursthouse provides a useful overview of the re-emergence of virtue ethics in recent decades, along with discussion of current approaches and issues.]

Thomson, Anne. 1999. *Critical Reasoning in Ethics: A Practical Introduction.* London: Routledge.
[Provides one of the most accessible and genuinely practical approaches to logic and critical thinking in ethics for the non-specialist (meaning, first of all, students and faculty outside the field of philosophy) that I have seen.]

Yu, Jiyuan. 2007. *The Ethics of Confucius and Aristotle: Mirrors of Virtue.* New York: Routledge.
[Provides one of the most complete comparisons and contrasts between these two major forms and traditions of virtue ethics.]

Information and computer ethics (ICE)

Because computers and then computer networks were among the earliest digital technologies, much of the extant reflection and literature on ethics and digital media is constituted by information and computing ethics. The literature, accordingly, is huge. Here are simply a few recommendations for further reading:

Himma, Kenneth Einar and Herman T. Tavani (eds.) 2008. *The Handbook of Information and Computer Ethics.* Hoboken, N.J.: John Wiley and Sons.
[An authoritative survey of foundational issues, methodological frameworks, and their applications in key foci (e.g., privacy,

intellectual property, security), along with major sections on professional ethics, responsibility issues and risk assessment, regulatory issues, and access and equity issues.]

Spinello, Richard A. and Herman T. Tavani (eds) 2004. *Readings in Cyberethics* (2nd ed.). Sudbury, MA: Jones and Bartlett.
[A comprehensive collective of primary readings on computer ethics and specific topics, including freedom of expression, intellectual property, privacy, and security and crime.]

Tavani, Herman T. 2007. *Ethics and Technology: Ethical Issues in an Age of Information and Communication Technology* (2nd ed). Hoboken, N.J.: John Wiley and Sons.
[One of the single most comprehensive and clear books in this domain, especially useful for teaching as it includes an extensive array of real-world case-studies.]

van den Hoven, Jeroen and Weckert, John (eds.). 2008. *Information Technology and Moral Philosophy*. Cambridge: Cambridge University Press.
[An essential collection of philosophically sophisticated analyses of a wide range of issues and topics in ICE by some of its most significant and well-known figures. Nicely complements the Himma and Tavani *Handbook* by providing more specialized articles.]

Ethical relativism and pluralism

Brey, Philip. 2007. Is Information Ethics Culture-Relative? *Journal of Technology and Human Interaction* 3 (3), 12–24.
[Brey – also known for his development of an approach to information ethics he denotes as "disclosive ethics" – works through in considerable detail how (and if so, how far) we may move beyond ethical and cultural relativism.]

Capurro, Rafael. 2008. Intercultural Information Ethics. In Kenneth Einar Himma and Herman T. Tavani (eds.), *The Handbook of Information and Computer Ethics*, 639–65. Hoboken, N.J.: John Wiley and Sons.
[Capurro coined the term "intercultural information ethics" (IIE) and here provides a masterful and critical overview of the growing field of literature, topics, traditions, and approaches. To my knowledge, the best single introduction to IIE in print.]

Ess, Charles. 2007b. Cybernetic Pluralism in an Emerging Global Information and Computing Ethics. *International Review of Information Ethics*, 7 (September). <http://www.i-r-i-e.net/inhalt/007/11-ess.pdf>, accessed September 23, 2008.

[An extensive review of ethical pluralism in information and computing ethics, including more detailed attention to both Confucian and African contexts.]

(see also: ———. 2007a. Bridging Cultures: Theoretical and Practical Approaches to Unity and Diversity Online. (Introduction to Special Issue, Information Ethics.) *International Journal of Technology and Human Interaction 3* (3), iii–x.)

Floridi, Luciano. 2007. Global Information Ethics: The Importance of Being Environmentally Earnest. *Journal of Technology and Human Interaction 3* (3), 1–11.

[Floridi – the author of one of the most widely influential frameworks for information and computing ethics, namely, his ontological theory of information ethics – takes up here the specific challenges of cultural diversity to a global ICE. He argues specifically for an ethical pluralism in the form of what he calls a "lite" information ontology.]

Hiruta, Kei. 2006. What Pluralism, Why Pluralism, and How? A Response to Charles Ess. *Ethics and Information Technology 8* (4), 227–36.

[Hiruta responds to my earlier effort to sketch out a particular version of ethical pluralism – namely, a *pros hen* or focal pluralism as based on Aristotle (Ess 2006c). While generally sympathetic with the project, Hiruta helpfully notes some important weaknesses and potential alternatives.]

Stahl, Bernd Carsten. 2006. Emancipation in Cross-Cultural IS [Information Systems] Research: The Fine Line between Relativism and Dictatorship of the Intellectual. *Ethics and Information Technology 8* (3), 97–108.

[Stahl (2004) has developed his own framework for information ethics in cross-cultural contexts, one resting on especially the procedural approaches highlighted in the work of Jürgen Habermas. Here, Stahl explores ways of avoiding cultural imperialism while nonetheless advocating emancipation as a central ethical goal and norm. Stahl's article is usefully read in conjunction with Wheeler below.]

Wheeler, Deborah. 2006. Gender sensitivity and the drive for IT: Lessons from the NetCorps Jordan Project. *Ethics and Information Technology 8* (3), 131–42.

[Wheeler is one of the most prolific ethnographers of women and IT in the Middle East. Here she highlights an approach to an ICT4D (ICT for Development) project in Jordan, and its differential impacts on women and men. Arguably, the introduction of IT in this project has led to a form of emancipation for women in Jordan, but one that is understood pluralistically, i.e., one that retains and preserves some of the essential differences defining local culture(s) in particular and Arab-Islamic cultures more broadly.]

Feminist approaches to digital media ethics

Adam, Alison. 2005. *Gender, Ethics and Information Technology*. New York: Palgrave MacMillan.
[Includes introductory chapters on ethics, computer ethics, and issues of gender – and then applies these to problems of "Internet Dating, Cyberstalking and Internet Pornography" (ch. 6), "Hacking into Hacking: Gender and the Hacker Phenomenon" (ch. 7), and "Someone to Watch Over me – Gender, Technologies, Privacy and Surveillance" (ch. 8).]
——. 2008. The Gender Agenda in Computer Ethics. In Kenneth E. Himma and Herman T. Tavani (eds.), *The Handbook of Information and Computer Ethics*, 589–619. Hoboken, N.J.: John Wiley and Sons.
[Provides a critical review of current accounts of women's and men's decision-making, underlining the need for better theorizing concerning gender. Adam highlights cyberstalking and hacker ethics as important areas for feminist analysis and points towards a cyberfeminism that conjoins political goals, subversion, and playfulness as dimensions of feminist ethics and the ethics of care that would help further develop a feminist computer ethics.]
van der Velden, Maya. Forthcoming. Design for a Common World: On Ethical Agency and Cognitive Justice. *Ethics and Information Technology* DOI: 10.1007/s10676-008-9178-2.
[Uses a feminist critique of positivist understandings of technology and science as ethically neutral, in order to develop an alternative notion of moral agency as distributed among interconnected designers, the technologies they design, and the users of these technologies.]

Advanced topics in ethical theory: post-structuralism and postmodernism

In addition to recent developments in virtue theory and feminist ethics, ethical theory in the West has undergone dramatic
re-examination under the influence of the post-structuralist

and postmodernist movements that began in the 1970s. While it appears that the influence – sometimes highly relativistic – of thinkers in these movements is on the wane, additional work in media ethics (or any other ethics, for that matter) must take their key figures and arguments into account.

One possible gateway into these domains is:

Braidotti, Rosa. 2006. *Transpositions: On Nomadic Ethics.* Cambridge: Polity Press.[3]
[Braidotti offers a novel ethical framework, in part based in Deleuze, that seeks to move beyond the basic assumptions of modern Western ethics – including the primary assumption of a singular *identity* – while also countering standard critiques of such postmodernist approaches, e.g., that their anti-universalism appears to run the risk of ethical relativism.]

Advanced topics in applied ethics: developing a digital media ethics on the basis of African thought

The first African Information Ethics conference in 2007, discussed above, has issued in a significant collection of essays and reflections – from both within more prevailing Western frameworks *and* within distinctively African perspectives.

Review the papers now available on the website of the *International Review of Information Ethics* for Volume 7. It is strongly recommended that you begin with Rafael Capurro's opening address (2007), but in the listing for the first 15 papers you will no doubt see others that will also provide distinctive introductions to African frameworks and issues in ICE.

---

3 I am grateful to Dr. Randi Markussen, Department of Information and Media Studies, University of Aarhus, for making me aware of this source.

Glossary

Affirming the consequent: a common logical fallacy (invalid argument). Formulaically, the argument looks like this: A→B (If A is true, then B is true). B (B is true). Therefore A (A is true).

CMC: computer-mediated communication – e.g., email and listserves, chat, videoconferencing, etc., as these all depend upon both computers and computer networks to function.

Copyleft: a range of schemes for protecting author's rights to their work, but in ways that are less restrictive with regard to how others may take up such work, as compared with prevailing copyright frameworks. Primary examples of such copyleft schemes are the GNU Free Documentation License and the Creative Commons Licenses.

Cyber-bullying: using various communicative venues – e.g., message boards on a MySpace or Facebook profile, email, etc. – to attack and intimidate another person. (In a recent U.S. case, it is charged that such cyber-bullying led to the suicide of a 13-year-old girl, Megan Meier.)

Cyber-stalking: using the Internet and/or other digital media (e.g. "buddy-locators" on cellphones) to stalk or harass a person.

Exclusive **or**: "Either one or the other – but not both." More formulaically: Either A may be true or B may be true, but both A and B may not be true at the same time.

While the exclusive or describes many relationships properly (e.g., a standard light may be on, or off, but not both), when applied to contexts and examples that in fact admit of a third possibility (or more), the exclusive or then becomes a *false dichotomy*. Noted in this volume because of its prevalence in media reporting fostering "moral panics."

Flaming: sending especially aggressive, insulting, hostile messages, sometimes anonymously or under a pseudonym. Flaming sometimes results in "flame-wars" – intensively hostile exchanges between a few members of a listserv or other online group. Such flame-wars, in turn, can destroy online communities as they make further participation in the listserv too unpleasant and insufficiently rewarding to be justified.

FLOSS: Free/Libre/Open Source Software (see chapter 3, pp. 75–6).

F2F: "face-to-face": direct (face-to-face) conversation/communication, in contrast with communication mediated through such information and communication technologies as cellphones, text-messaging, email, etc.

ICTs: information and communication technologies.

Inclusive **or:** "Either one or the other – possibly both." More formulaically: Either A may be true or B may be true, or both A and B may be true at the same time.

Internet: the Internet (perhaps more accurately, a global collection of internets) is made up of computer networks based upon a standardized collection of machine communication protocols – clustered around the TCP (Transmission Control Protocol)/IP (Internet Protocol) suite.

The World Wide Web – familiar to most users in the form of webpages accessed through browsers such as Windows Explorer, Mozilla's Firefox, and others – may be considered a subset of the Internet, one built upon HTML (Hypertext Mark-up Language) coding that facilitates moving quickly from any one page to another (wherever such pages may be located on a given computer/server) in the form of links.

PDA: personal digital assistant (e.g., a Palm or other brand of hand-held computer, used primarily as an organizer, calendar, address book, etc.).

Questionable analogy: an analogy argument compares two (or more) examples/cases/contexts in order to conclude that what is accepted as true in a first example/case/context should be true of the second, because of the relevant *similarities* between the two. But if, upon analysis, one or more relevant *differences* between the two cases can be pointed out, then the analogy is *questionable* and the conclusion is not warranted. More idiomatically thought of as "comparing apples and oranges."

Trolling: a troll joins a listserv or other online community with the intention of provoking attention to himself or herself, her ideas, his favorite theories, etc. – and thereby taking over the communicative bandwidth of the group. Trolls may not always intend to be destructive, but they are infamously so, especially for groups that are comparatively vulnerable and/or populated by relatively new users of CMC. See Herring et al. (n.d.).

World Wide Web: see "Internet."

References

Abdat, Sjarif and Graham P. Pervan. 2000. Reducing the Negative Effects of Power Distance during Asynchronous Pre-Meeting with Using Anonymity in Indonesian Culture. In Fay Sudweeks and Charles Ess (eds.), *Proceedings of the Second International Conference on Cultural Attitudes towards Technology and Communication*, 209–15. Murdoch, Western Australia: Murdoch University Press.

Adam, Alison. 2005. *Gender, Ethics and Information Technology*. New York: Palgrave Macmillan.

——. 2008. The Gender Agenda in Computer Ethics. In Kenneth Einar Himma and Herman T. Tavani (eds.), *The Handbook of Information and Computer Ethics*, 589–619. Hoboken, N.J.: John Wiley and Sons.

Adams, Carol J. [1990] 2000. *The Sexual Politics of Meat: A Feminist-Vegetarian Critical Theory*. New York: Continuum.

——. 1996. "This Is Not Our Fathers' Pornography": Sex, Lies, and Computers. In Charles Ess (ed.), *Philosophical Perspectives on Computer-Mediated Communication*, 147–70. Albany, N.Y.: State University of New York Press.

Alemán, Carolos Galvan. 2008. Communication Modes, Hispanic. In Wolfgang Donsbach (ed.), *International Encyclopedia of Communication*. <http://www.communicationencyclopedia.com/public/>, accessed September 19, 2008.

Ames, Roger and Henry Rosemont, Jr. 1998. *The Analects of Confucius: A Philosophical Translation*. New York: Ballantine Books.

Annis, Joseph. 2006. Reports of Council on Medical Service. (American Medical Association). <http://www.ama-assn.org/ama1/pub/upload/mm/38/a-06cms.pdf>, accessed April 8, 2007.

Asia-Pacific Development Information Programme (APDIP). n.d. Case Studies on Free and Open Source Software. <http://www.apdip.net/resources/case/foss/>, accessed January 20, 2008.

Association for Computing Machinery (ACM). 1992. Code of Ethics. <http://www.acm.org/about/code-of-ethics>, accessed September 23, 2008.

Association of Internet Researchers (AoIR) Ethics Working Group. 2002. *Ethical Decision-Making and Internet Research: Recommendations from the AoIR Ethics Working Committee.* <http://www.aoir.org/reports/ ethics.pdf>, accessed September 23, 2008.

Baldwin, John R. 2008. Communication Modes, Western. In Wolfgang Donsbach (ed.), *International Encyclopedia of Communication.* <http:// www.communicationencyclopedia.com/public/>, accessed September 19, 2008.

Barber, Trudy. 2004. A Pleasure Prophecy: Predictions for the Sex Tourist of the Future. In Dennis D. Waskul (ed.), *net.seXXX: Readings on Sex, Pornography, and the Internet,* 322–36. New York: Peter Lang.

Barnard, Anne. 2008. MySpace Agrees to Lead Fight to Stop Sex Predators. *New York Times,* January 15. <http://www.nytimes.com/ 2008/01/15/us/us/15myspace.html>, accessed January 16, 2008.

BBC News. 2008a.Warning on Stealthy Windows Virus (January 11). <http://news.bbc.co.uk/2/hi/technology/7183008.stm>, accessed September 13, 2008.

BBC News. 2008b Timeline: Child Benefits Records Loss (June 25). <http://news.bbc.co.uk/1/hi/uk_politics/7104368.stm>, accessed September 23, 2008

Bizer, Johann. 2003. Grundrechte im Netz: Von der freien Meinungsäußerung bis zum Recht auf Eigentum. In Christiane Schulzki-Haddouti (Ed.), *Bürgerrechte im Netz,* 21–9. Bonn: Bundeszentrale für politische Bildung. Available online: <http:// www.bpb.de/files/FPQOF9.pdf>, accessed September 23, 2008.

Bogost, Ian. 2007. *Persuasive Games: The Expressive Power of Videogames.* Cambridge, MA: MIT Press.

Boss, Judith. 2005. *Analyzing Moral Issues* (3rd ed.). Boston: McGraw-Hill.

Bowell, Tracy and Gary Kemp. 2005. *Critical Thinking: A Concise Guide* (2nd ed.). London: Routledge

Braidotti, Rosa. 2006. *Transpositions: On Nomadic Ethics.* Cambridge: Polity Press.

Brey. Philip. 2007. Is Information Ethics Culture-Relative? *Journal of Technology and Human Interaction 3* (3), 12–24.

Briggs, Asa and Peter Burke. 2005. *A Social History of the Media* (2nd ed.). Cambridge, Malden, MA: Polity Press.

Britz, Johannes. 2007. The Internet: The Missing Link between the Information Rich and the Information Poor? In Rafael Capurro, Johannes Frühbauer, and Thomas Hausmanninger (eds.), *Localizing the Internet: Ethical Aspects in Intercultural Perspective,* 265–77. Munich: Fink Verlag.

Buchanan, Elizabeth and Charles Ess. 2005. The Ethics of e-Games. (Introduction to special issue.) *International Review of Information Ethics*, December. <http://www.i-r-i-e.net/issue4.htm>, accessed September 23, 2008.

Buchanan, Elizabeth and Kathrine Andrews Henderson. 2008. *Case Studies in Library and Information Science Ethics*. Jefferson, N.C.: McFarland.

Burk, Dan. 2007. Privacy and Property in the Global Datasphere. In S. Hongladarom and C. Ess (eds.), *Information Technology Ethics: Cultural Perspectives*, 94–107. Hershey, PA: IGI Global.

Bynum, Terrell Ward. 2000. A Very Short History of Computer Ethics. American Philosophical Association's *Newsletter on Philosophy and Computing*. <http://www.southernct.edu/organizations/rccs/resources/research/introduction/bynum_shrt_hist.html>, accessed September 23, 2008.

Calvert, Sandra. 2005. Cognitive Effects of Video Games. In Joos Raessen and Jeffrey Goldstein (eds.), *Handbook of Computer Game Studies*, 125–31. Cambridge, MA: MIT Press.

Canellopoulou-Bottis, Maria and Kenneth Einar Himma. 2008. The Digital Divide: A Perspective for the Future. In Kenneth Einar Himma and Herman T. Tavani (eds.), *The Handbook of Information and Computer Ethics*, 621–37. Hoboken, N.J.: John Wiley and Sons.

Capurro, Rafael. 2005. Privacy: An Intercultural Perspective. *Ethics and Information Technology* 7 (1), 37–47.

——. 2007. Information Ethics for and from Africa. *IRIE International Review of Information Ethics* 7 (09). <http://www.i-r-i-e.net/inhalt/007/01-capurro.pdf>, accessed September 19, 2008.

——. 2008. Intercultural Information Ethics. In Kenneth Einar Himma and Herman T. Tavani (eds.), *The Handbook of Information and Computer Ethics*, 639–65. Hoboken, N.J.: John Wiley and Sons.

Cavalier, Robert. 1996. Feminism and Pornography: A Dialogical Perspective. <http://caae.phil.cmu.edu/Cavalier/Forum/pornography/background/CMC_article.html>, accessed January 18, 2008.

Cavalier, Robert and Charles Ess (eds.). 1996. Academic Dialogue on Applied Ethics: Feminists' Perspectives on Pornography. <http://caae.phil.cmu.edu/Cavalier/Forum/pornography/porn.html>, accessed January 18, 2008.

Chan, Joseph. (2003). Confucian Attitudes towards Ethical Pluralism. In Richard Madsen and Tracy B. Strong (eds.), *The Many and the One: Religious and Secular Perspectives on Ethical Pluralism in the Modern World*, 129–53. Princeton: Princeton University Press.

Chokvasin, Theptawee. 2007. Mobile Phone and Autonomy. In Soraj
    Hongladarom and Charles Ess (eds.), *Information Technology Ethics:
    Cultural Perspectives*, 68–80. Hershey, PA: Idea Group Reference.
Chopra, Samir and Scott Dexter. 2008. *Decoding Liberation: The Promises
    of Free and Open Source Software*. New York: Routledge.
Christoffersen, Lisbet (ed.). 2006. *Gudebilleder – Ytringsfrihed og religion i
    en globaliseret verden* [*Images of God: Freedom of Speech and Religion in a
    Globalized World*]. Copenhagen: Tiderne Skifter.
Coleman, Kari Gwen. 2001. Android Arête: Toward a Virtue Ethic for
    Computational Agents. *Ethics and Information Technology 3* (4), 247–65.
Consalvo, Mia. 2005. Rule Sets, Cheating, and Magic Circles: Studying
    Games and Ethics. *International Review of Information Ethics 4*
    (December), 7–12. <http://www.i-r-i-e.net/inhalt/004/Consalvo.pdf>,
    accessed September 23, 2008.
——. 2007. *Cheating: Gaining Advantage in Videogames*. Cambridge, MA:
    MIT Press.
Constitution of the People's Republic of China. 1982. <http://en.
    chinacourt.org/public/detail.php?id=2697>, accessed December 10,
    2007.
Council of Europe. 1950. The European Convention on Human Rights
    and Its Five Protocols. <http://www.hri.org/docs/ECHR50.html>,
    accessed December 10, 2007.
Critcher, Chas. 2006. *Critical Readings: Moral Panics and the Media*. New
    York: McGraw-Hill.
Debatin, Bernhard (ed.). 2007. *The Cartoon Debate and the Freedom of the
    Press. Conflicting Norms and Values in the Global Media Culture/Der
    Karikaturenstreit und die Pressefreiheit. Wert- und Normenkonflikte in der
    globalen Medienkultur*. Berlin: LIT Verlag.
Descartes, René. [1637] 1972. *Discourse on Method*. In *The Philosophical
    Works of Descartes*, E.S. Haldane and G.R.T. Ross (trans.), Volume I,
    81–130. Cambridge: Cambridge University Press.
——. [1641] 1972. *Meditations on First Philosophy*. In *The Philosophical
    Works of Descartes*, E.S. Haldane and G.R.T. Ross (trans.), Volume I,
    135–99. Cambridge: Cambridge University Press.
Die Welt. 2007. Die Kinder-Ermittler der Ministerin [The [Federal Family]
    Minister's Children-Investigators], October 13, 1. <http://
    www.welt.de/welt_print/article1261600/Die_Kinder-
    Ermittler_der_Ministerin.html>, accessed January 16, 2008
Durkin, Keith F. 2004. The Internet as a Milieu for the Management of a
    Stigmatized Sexual Identity. In Dennis D. Waskul (ed.), *net.seXXX.
    Readings on Sex, Pornography, and the Internet*, 131–47. New York: Peter
    Lang.

Dyson, Laurel, Max Hendriks, and Stephen Grant (eds.). 2007. *Information Technology and Indigenous People*. Hershey, PA: Information Science Publishing.

Eickelman, Dale F. (2003). Islam and Ethical Pluralism. In Richard Madsen and Tracy B. Strong (eds.), *The Many and the One: Religious and Secular Perspectives on Ethical Pluralism in the Modern World*, 161–80. Princeton: Princeton University Press.

Ess, Charles. 1998. Cosmopolitan Ideal or Cybercentrism? A Critical Examination of the Underlying Assumptions of "The Electronic Global Village." *APA Newsletter on Computers* 97 (2: Spring). <http://www.apa.udel.edu/apa/archive/newsletters/v97n2/computers/ess.asp>, accessed October 4, 2008.

——. 2005. "Lost in Translation"? Intercultural Dialogues on Privacy and Information Ethics. (Introduction to Special Issue on Privacy and Data Privacy Protection in Asia.) *Ethics and Information Technology* 7 (1), 1–6.

——. 2006a. Du colonialisme informatique à un usage culturellement informé des TIC [From Computer-Mediated Colonization to Culturally Aware ICT Usage and Design]. In Joëlle Aden (ed.), *De Babel à la mondialisation: apport des sciences sociales à la didactique des langues* [*From Babel to Globalization: The Contribution of Social Sciences to Language Teaching*], 47–61. Dijon: CNDP – CRDP de Bourgogne.

——. 2006b. From Computer-Mediated Colonization to Culturally-Aware ICT Usage and Design. In Panayiotis Zaphiris and Sri Kurniawan (eds.), *Advances in Universal Web Design and Evaluation: Research, Trends and Opportunities*, 178–97. Hershey, PA: Idea Publishing.

——. 2006c. Ethical Pluralism and Global Information Ethics. *Ethics and Information Technology* 8 (4), 215–26.

——. 2006d. Universal Information Ethics? Ethical Pluralism and Social Justice. In Emma Rooksby and John Weckert (eds.), *Information Technology and Social Justice*, 69–92. Hershey, PA: Idea Publishing.

——. 2007a. Bridging Cultures: Theoretical and Practical Approaches to Unity and Diversity Online. (Introduction to Special Issue, Information Ethics.) *International Journal of Technology and Human Interaction* 3 (3), iii–x.

——. 2007b. Cybernetic Pluralism in an Emerging Global Information and Computing Ethics. *International Review of Information Ethics* 7 (September). <http://www.i-r-i-e.net/inhalt/007/11-ess.pdf>, accessed September 23, 2008.

——. 2007c. Déclinaisons culturelles en ligne: observation «de l'autre» [Cultural Declensions Online: Watching Out for "the Other"]. Special issue of *Études de Linguistique Appliquée* [*Studies of Applied Linguistics*].

Paris: Didier Klienkensick. *D'autres espaces pour les cultures*
[*Other Spaces for Cultures*], Clara Farrao, ed. No. 146 (avril–juin),
149–60.

———. n.d. The Letter from the Birmingham Jail: Critical Thinking Notes.
<http://www.drury.edu/ess/alpha/mlking.html>, accessed September
14, 2008.

Ess, Charles and Fay Sudweeks. 2001. On the Edge: Cultural Barriers and
Catalysts to IT Diffusion among Remote and Marginalized
Communities. *New Media and Society 3* (3), 259–69.

———. 2005. Culture and Computer-Mediated Communication: Toward
New Understandings. *Journal of Computer-Mediated Communication*
11 (1), article 9. <http://jcmc.indiana.edu/vol11/issue1/ess.html>,
accessed September 19, 2008.

European Union. 1995. DIRECTIVE 95/46/EC OF THE EUROPEAN
PARLIAMENT AND OF THE COUNCIL of 24 October 1995. <http:-
//eur-lex.europa.eu/LexUriServ/LexUriServ.do?uri=CELEX:31995L00
46:EN:HTML>, accessed October 4, 2008.

Finnemann, Niels Øle. 2005. *Internettet i mediehistorisk perspektiv [The
Internet in Media-Historical Perspective]*. Frederiksberg, Denmark:
Samfundslitteratur.

Finquelievich, Susana. 2007. A Toolkit to Empower Communities in Latin
America. In Rafael Capurro, Johannes Frühbauer, and Thomas
Hausmanninger (eds.), *Localizing the Internet: Ethical Aspects in
Intercultural Perspective*, 301–19. Munich: Fink Verlag.

Floridi, Luciano. 2005. The Ontological Interpretation of Informational
Privacy. *Ethics and Information Technology 7* (4), 185–200.

———. 2006. Four Challenges for a Theory of Informational Privacy. *Ethics
and Information Technology 8* (3), 109–19.

———. 2007. Global Information Ethics: The Importance of Being
Environmentally Earnest. *Journal of Technology and Human Interaction
3* (3), 1–11.

———. 2008. Information Ethics: A Reappraisal. *Ethics and Information
Technology 10* (2–3), 189–204.

Franklin, Benjamin. [1784]. *The Autobiography of Benjamin Franklin*,
Chapter Eight. <http://www.earlyamerica.com/lives/franklin/
chapt8/>, accessed October 4, 2008.

FUNREDES. n.d. *Fundación-Redes-y-Desarrollo* / Networks-and-
Development-Foundation/*Association-Réseaux-et-Développement*.
<http://www.funredes.org/english/index.php3>, accessed January 20,
2008.

Gazzaniga, Michael S. 2005. *The Ethical Brain*. Washington, D.C.: Dana
Press.

Gatterburg, Angela. 2007. Aliens im Kinderzimmer [Aliens in the Kids' Room]. *Der Spiegel* 20 (May 14), 42–54.

Gee, James Paul. 2003. *What Video Games Have to Teach Us about Learning and Literacy*. New York: Palgrave Macmillan.

Gelbrich, Katja and Holger Roschk. 2008. Advertising, Cross-Cultural. In Wolfgang Donsbach (ed.), *International Encyclopedia of Communication*. <http://www.communicationencyclopedia.com/public/>, accessed September 19, 2008.

Ghosh, Shohini. 2006. The Troubled Existence of Sex and Sexuality: Feminists Engage with Censorship. In Christiane Brosius and Melissa Butcher (eds.), *Image Journeys: Audio-Visual Media and Cultural Change in India*, 233–60. Delhi: Sage.

Gilligan, Carol. 1982. *In a Different Voice: Psychological Theory and Women's Development*. Cambridge, MA: Harvard University Press.

Gimmler, Antje.1996. The Discourse Ethics of Jürgen Habermas. <http://caae.phil.cmu.edu/cavalier/Forum/meta/background/agimmler.html>, accessed July 16, 2008.

Gordon, Wendy J. 2008. Moral Philosophy, Information Technology, and Copyright: The Grokster Case. In Jeroen van den Hoven and John Weckert (eds.), *Information Technology and Moral Philosophy*, 270–300. Cambridge: Cambridge University Press.

Grodzinsky, Frances S. and Marty Wolf. 2008. Ethical Interest in Free and Open Source Software. In Kenneth Einar Himma and Herman T. Tavani (eds.), *The Handbook of Information and Computer Ethics*, 245–71. Hoboken, N.J.: John Wiley and Sons.

GVU (Graphic, Visualization, and Usability Center, Georgia Technological University). 1998. GVU's 10th WWW User Survey. <http://www-static.cc.gatech.edu/gvu/user_surveys/survey-1998-10/graphs/general/q50.htm>, accessed January 19, 2008.

Hall, Edward T. 1976. *Beyond Culture*. New York: Anchor Books.

Hashim, Noor Hazarina, Jamie Murphy, and Nazlida Muhamad Hashim. 2007. Islam and Online Imagery on Malaysian Tourist Destination Websites. *Journal of Computer-Mediated Communication*, 12 (3), article 16. <http://jcmc.indiana.edu/vol12/issue3/hashim.html>, accessed December 11, 2007.

Hausmanninger, Thomas. 2007. Allowing for Difference: Some Preliminary Remarks Concerning Intercultural Information Ethics. In Rafael Capurro, Johannes Frühbauer, and Thomas Hausmanninger (eds.), *Localizing the Internet: Ethical Aspects in Intercultural Perspective*, 39–56. Munich: Fink Verlag.

Heaton, Lorna. 2001. Preserving Communication Context: Virtual Workspace and Interpersonal Space in Japanese CSCW. In Charles Ess (ed.), *Culture, Technology, Communication: Towards an Intercultural Global Village*, 213–40. Albany, N.Y.: State University of New York Press.

Hecht, Michael L. 2008. Communication Modes, African. In Wolfgang Donsbach (ed.), *International Encyclopedia of Communication*. <http://www.communicationencyclopedia.com/public/>, accessed September 19, 2008.

Hermeking, Marc. 2005. Culture and Internet Consumption: Contributions from Cross-Cultural Marketing and Advertising Research. *Journal of Computer-Mediated Communication, 11* (1), article 10. <http://jcmc.indiana.edu/vol11/issue1/hermeking.html>, accessed September 23, 2008.

Herring, Susan, Kirk Job-Sluder, and Sasha Barab. n.d. Searching for Safety Online: Managing "Trolling" in a Feminist Forum. *The Information Society 18* (5), 371–83.

Hewling, Anne. 2005. Culture in the Online Class: Using Message Analysis to Look Beyond Nationality-Based Frames of Reference. *Journal of Computer-Mediated Communication, 11* (1), article 16. <http://jcmc.indiana.edu/vol11/issue1/hewling.html>, accessed September 30, 2008.

Himma, Kenneth Einar. 2007. The Information Gap, the Digital Divide, and the Obligations of Affluent Nations. *IRIE International Review of Information Ethics, 7* (09). <http://www.i-r-i-e.net/inhalt/007/07-himma.pdf>, accessed September 19. 2008.

——. 2008. The Justification of Intellectual Property: Contemporary Philosophical Disputes. *Journal of the American Society for Information Sciences and Technology 59* (7), 1143–61.

Hinman, Lawrence M. 2008. *Ethics: A Pluralistic Approach to Moral Theory* (4th ed.). Belmont, CA: Wadsworth.

Hiruta, Kei. 2006. What Pluralism, Why Pluralism, and How? A Response to Charles Ess. *Ethics and Information Technology 8* (4), 227–36.

Hofstede, Gert. 1980. *Culture's Consequences: International Differences in Work-Related Values*. Beverly Hills, CA: Sage.

——. 1983. National Cultures in Four Dimensions. *International Studies of Management and Organization 13*, 52–60.

——. 1984. The Cultural Relativity of the Quality of Life Concept. *Academy of Management Review 9*, 389–98.

——. 1991. *Cultures and Organizations: Software of the Mind*. London: McGraw-Hill.

Hongladarom, Soraj. 2007a. Analysis and Justification of Privacy from a Buddhist Perspective. In Soraj Hongladarom and Charles Ess (eds.), *Information Technology Ethics: Cultural Perspectives*, 108–22. Hershey, PA: Idea Group Reference.

——. 2007b. Information Divide, Information Flow and Global Justice. *IRIE International Review of Information Ethics* 7 (09). <http://www.i-r-i-e.net/inhalt/007/08-hongladarom.pdf>, accessed September 19, 2008.

Hongladarom, Soraj and Charles Ess. 2007. Preface. In Soraj Hongaldarom and Charles Ess (eds.), *Information Technology Ethics: Cultural Perspectives*, xi–xxxv. Hershey, PA: Idea Reference.

Hughes, Donna. 2004. The Use of New Communications and Information Technologies for Sexual Exploitation of Women and Children. In Dennis D. Waskul (ed.), *net.seXXX: Readings on Sex, Pornography, and the Internet*, 109–30. New York: Peter Lang.

Huizinga, Johan. [1938] 1955. *Homo Ludens: A Study of the Play-Element in Culture*. Boston: Beacon Press.

Hursthouse, Rosalind. 1999. *On Virtue Ethics*. Oxford: Oxford University Press.

Jackson, Willy and Issiaka Mandé. 2007. "New Technologies" and "Ancient Africa": The Impact of New Information and Communication Technologies in Sub-Saharian [*sic*] Africa. In Rafael Capurro, Johannes Frühbauer, and Thomas Hausmanninger (eds.), *Localizing the Internet: Ethical Aspects in Intercultural Perspective*, 171–6. Munich: Fink Verlag.

Jacobs, Katrien, Marije Janssen, and Matteo Pasquinelli (eds.). 2007. *C'Lick Me: A Netporn Studies Reader*. Amsterdam: Institute of Network Cultures. Available online: <http://networkcultures.org/wpmu/portal/publications/inc-readers/the-art-and-politics-of-netporn/>, accessed October 2, 2008.

Jefferson, Thomas. [1776] 1984. A Declaration by the Representatives of the United States of America, in General Congress Assembled. In Merrill D. Peterson (ed.), *Thomas Jefferson: Writings*, 19–24. New York: Library of America.

Jenkins, Henry. 2006. *Convergence Culture: Where Old and New Media Collide*. New York: New York University Press.

Jesdanun, Anick. 2007. Filtering: Etiquette Questions Arise as Airlines Introduce In-Flight Internet Access. Associated Press. <http://www.chicagotribune.com/travel/chi-ap-airborne-webdec24,0,2805500.story>, accessed January 18, 2008.

Johnson, Deborah. 2001. *Computer Ethics* (3rd ed.). Upper Saddle River, N.J.: Prentice-Hall.

Kahlweit, Cathrin. 2007. Happy Slapping und Snuff-Videos: Bilder von enormer Schlagkraft [Happy Slapping and Snuff Videos: Images with Enormous Impact]. *Suddeutsche Zeitung*, January 31. <http://www.

sueddeutsche.de/deutschland/artikel/107/100007/>, accessed September 23, 2008.

Kampf, Constance. 2008. From Culture to Cultural Attitudes, Knowledge Communication Practices and Innovation. In Fay Sudweeks, Herbert Hrachovec, and Charles Ess (eds.), *Proceedings: Cultural Attitudes towards Communication and Technology 2008*, 581–91. School of Information Technology, Murdoch University: Murdoch, Western Australia.

Kant, Immanuel. [1788] 1956. *Critique of Practical Reason*. Lewis White Beck, trans. Indianapolis: Bobbs-Merrill.

———. [1785] 1959. *Foundations of the Metaphysics of Morals*. Lewis White Beck, trans.. Indianapolis: Bobbs-Merrill.

King, Martin Luther King, Jr. [1963] 1964. Letter from the Birmingham Jail. In Martin Luther King, Jr. (ed.), *Why We Can't Wait*, 77–100. New York: Mentor.

Kitiyadisai, Krisana. 2005. Privacy Rights and Protection: Foreign Values in Modern Thai Context. *Ethics and Information Technology* 7 (1), 17–26.

Kunelius, Risto, Elisabeth Eide, Oliver Hahn, and Roland Schroeder. 2007. *Reading the Mohammed Cartoons Controversy: An International Analysis of Press Discourses on Free Speech and Political Spin*. Bochum, Germany: Projekt Verlag

Lenhart, Amanda and Mary Madden. 2007. *Teens, Privacy and Online Social Networks: How Teens Manage Their Online Identities and Personal Information in the Age of MySpace*. Washington, D.C.: Pew Internet and American Life Project. Available from <http://www.pewinternet.org/PPF/r/211/report_display.asp>, accessed January 20, 2008.

Levinas, Emmanuel. 1963. La Trace de l'autre [The Trace of the Other]. Alphonso Lingis, trans. *Tijdschrift voor Philosophie* (Sept.), 605–23.

———. 1987. *Time and the Other and Additional Essays*. Richard A. Cohen, trans. Pittsburgh, PA: Duquesne University Press.

Lieberman, Debra. 2006. What Can We Learn from Playing Interactive Games? In Peter Vorderer and Jennings Bryant, J. (eds.), *Playing video games: Motives, responses, and consequences*, 379–97. Mahwah, N.J.: Lawrence Erlbaum Associates.

Lievrouw, Leah A. and Sonia Livingstone (eds.). 2006. *The Handbook of New Media: Social Shaping and Social Consequences of ICTs (Updated Student Edition)*. Thousand Oaks, CA: Sage Publications.

Lim, Merlyna. 2006. Democracy, Conspiracy, Pornography: The Internet and Political Activism in Indonesia. Lecture at IR 7.0: Internet Convergences Conference, Brisbane, September 28.

Lin, Pei-Chun and Caroline Henkes. 2004. Privacy Research: Privacy Sense in China, Hong Kong and Taiwan. Unpublished MS: Universität Trier.

Lü, Yao-Hui. 2005. Privacy and Data Privacy Issues in Contemporary China. *Ethics and Information Technology* 7 (1), 7–15.

Macfadyen, Leah P. 2008. The Perils of Parsimony: "National Culture" as Red Herring? In Fay Sudweeks, Herbert Hrachovec and Charles Ess (eds.), *Proceedings Cultural Attitudes towards Communication and Technology 2008*, 569–80. School of Information Technology, Murdoch University: Murdoch, Western Australia.

McFarland, Michael C. 2004. Intellectual Property, Information, and the Common Good. In Richard A. Spinello and Herman T. Tavani (eds.), *Readings in CyberEthics* (2nd ed.), 294–304. Sudbury, MA: Jones and Bartlett.

MacKinnon, Catharine. 1989. *Toward a Feminist Theory of State.* Cambridge, MA: Harvard University Press.

——. 1992. Pornography, Civil Rights and Speech. In Catherine Itzin (ed.), *Pornography: Women, Violence and Civil Liberties: A Radical New View*, 456–511. Oxford: Oxford University Press.

——. 1993. *Only Words.* Cambridge, MA: Harvard University Press.

Mainz, Pernille. 2007. "Fråderen" og "luder" er nye ord på mobilen:Ekspert frygter, at det på sigt vil udvikle unge menneskers sprog i en negativ retning ["Foaming' {Slang for "Hungry"} and "Slut" are the New Words on Mobile Phones: An Expert Wonders Whether This Means in the Long Run That Young People's Language Will Develop in a Negative Direction]. *Nyhedsavisen*, December 15, 14.

Menkiti, Ifeanyi A. 1979. Person and Community in African Traditional Thought. In Richard A. Wright (ed.), *African Philosophy*, 157–68. New York: New York University Press.

Mesbahi, Mohamed. 2007. The Third World and the Paradox of the Digital Revolution. *IRIE International Review of Information Ethics* 7 (09). <http://www.i-r-i-e.net/inhalt/007/04-mesbahi.pdf>, accessed September 19, 2008.

Midgley, Mary. [1981] 1996. Trying Out One's New Sword. In *Heart and Mind*. New York: St. Martin's Press; reprinted in John Arthur (ed.), *Morality and Moral Controversies* (4th ed.), 116–19. Upper Saddle River, N.J.: Simon and Schuster.

Moghaddam, Fathali M. 2008. Communication Modes, Muslim. In Wolfgang Donsbach (ed.), *International Encyclopedia of Communication*. <http://www.communicationencyclopedia.com/public/>, accessed September 19, 2008.

Moor, James. 1997. Towards a Theory of Privacy in the Information Age. *ACM SIGCAS Computers and Society 27*, 27–32.

——. 2002. Toward a Theory of Privacy in the Information Age. In Robert M. Baird, Reagan Ramsower, and Stuart E. Rosenbaum (eds.), *Cyberethics: Social and Moral Issues in the Computer Age*, 200–12. Amherst, N.Y.: Prometheus Books.

Moore, Adam D. 2003. Privacy: Its Meaning and Value. *American Philosophical Quarterly 40* (3), 215–27.

——. 2008. Personality-Based, Rule-Utilitarian, and Lockean Justifications of Intellectual Property. In Kenneth Einar Himma and Herman T. Tavani (eds.), *The Handbook of Information and Computer Ethics*, 105–30. Hoboken, N.J.: John Wiley and Sons.

Mowlabocus, Sharif. 2007. Gay Men and the Pornification of Everyday Life. In Susanna Paasonen, Kaarina Nikunen, and Laura Saarenmaa Paasonen (eds.), *Pornification: Sex and Sexuality in Media Culture*, 61–71. Oxford, New York: Berg Publishers.

Mullins, Phil. 1996. Sacred Text in the Sea of Texts: The Bible in North American Electronic Culture. In Charles Ess (ed.), *Philosophical Perspectives on Computer-Mediated Communication*, 271–302. Albany, N.Y.: SUNY Press.

Myskja, Bjørn. 2008. The Categorical Imperative and the Ethics of Trust. *Ethics and Information Technology 10* (4), 213–20.

Nagenborg, Michael. 2005. *Das Private unter den Rahmenbedingungen der IuK-Technologie: Ein Beitrag zur Informationsethik [The Private Sphere under the Framing Conditions of Information and Communication Technologies: A Contribution to Information Ethics]*. Wiesbaden: Verlag für Sozialwissenschaften.

Nakada, Makoto and Takanori Tamura. 2005. Japanese Conceptions of Privacy: An Intercultural Perspective. *Ethics and Information Technology 7* (1), 27–36.

Networld. 2005. NRW-Jugendminister warnt vor Panikmache bei Computerspielen [North Rhein-Westphalia Youth Minister Warns Against Computer-Game Panic]. *Networld*, December 1. <http://www.golem.de/0512/41943.html>, accessed September 24, 2008.

Nikunen, Kaarina. 2007. *Cosmo* Girls Talk: Blurring Boundaries of Porn and Sex. In Susanna Paasonen, Kaarina Nikunen, and Laura Saarenmaa (eds.), *Pornification: Sex and Sexuality in Media Culture*, 73–85. Oxford, New York: Berg Publishers.

Nocera, José L. Abdelnour. 2008. Culture as a Bottom Up Concept to Understand Diversity in Systems Production and Use. In Fay Sudweeks, Herbert Hrachovec and Charles Ess (eds.), *Proceedings Cultural Attitudes Towards Communication and Technology 2008*,

540–50. School of Information Technology, Murdoch University: Murdoch, Western Australia.

Olsen, Kristian Leider. 2007a. Drenge filmer sex med intetanende piger [Boys Film Sex with Unsuspecting Girls]. *Nyhedsavisen*, November 29, 1.

——. 2007b. Sexvideoer skader piger [Sex Videos Hurt Girls]. *Nyhedsavisen*, November 29, 8.

Paasonen, Susanna. 2007. Netporn, Media Specificity and the Problem of the Mainstream. Presented at Internet Research 8.0 (Association of Internet Researchers) conference, Vancouver, B.C., October 19. Unpublished MS, cited by permission from the author.

Paasonen, Susanna, Kaarina Nikunen, and Laura Saarenmaa (eds.). 2007a. *Pornification: Sex and Sexuality in Media Culture*. Oxford, New York: Berg Publishers.

——. 2007b. Pornification and the Education of Desire (Introduction). In Susanna Paasonen, Kaarina Nikunen, and Laura Saarenmaa Paasonen (eds.), *Pornification: Sex and Sexuality in Media Culture*, 1–20. Oxford, New York: Berg Publishers.

Paterson, Barbara. 2007. We Cannot Eat Data: The Need for Computer Ethics to Address the Cultural and Ecological Impacts of Computing. In Soraj Hongladarom and Charles Ess (eds.), *Information Technology Ethics: Cultural Perspectives*, 153–68. Hershey, PA: Idea Reference.

Paul, Pamela. 2005. *Pornified: How Pornography Is Transforming Our Lives, Our Relationships, and Our Families*. New York: Time Books.

Perdue, Lewis. 2004. EroticaBiz: How Sex Shaped the Internet. In Dennis D. Waskul (ed.), *net.seXXX: Readings on Sex, Pornography, and the Internet*, 259–93. New York: Peter Lang.

Picturephoning.com n.d. Archives for the Category: Happy Slapping/Violence. <http://www.textually.org/picturephoning/archives/cat_happy_slappingviolence.htm>, accessed September 13, 2008.

Pimienta, Daniel. *Digital Divide, Social Divide, Paradigmatic Divide*. <http://funredes.org/mistica/english/cyberlibrary/thematic/Paradig matic_Divide.pdf> accessed January 20, 2008.

Plato. 1892. *The Apology* in *The Dialogues of Plato* (3rd ed.), vol. 2, 109–35. B. Jowett, trans. Oxford: Oxford University Press.

——. 1991. *The Republic*. Allan Bloom, trans., with notes, an interpretive essay and a new introduction. New York: Basic Books.

Pogue, David. 2007a. The Generational Divide in Copyright Morality. *New York Times*, December 20. <http://www.nytimes.com/2007/12/20/technology/personaltech/20pogue-email.html?8cir&emc=cir>, accessed September 24, 2008.

——. 2007b. Pogue's Posts: Readers Respond to the Debate over Responsible Downloading. *New York Times*, December 27. <http://pogue.blogs.nytimes.com/2007/12/27/readers-respond-to-the-debate-over-responsible-downloading/>, accessed September 24, 2008.

Possin, Kevin. 2005. *Critical Thinking: A Computer-Assisted Introduction to Logic and Critical Thinking* (CD). Winona, Minnesota: The Critical Thinking Lab.

Ramasoota, Pirongrong. 2001. Privacy: A Philosophical Sketch and a Search for a Thai Perception. *MANUSYA: Journal of Humanities 4* (2), 89–107.

Ramose, Mogobe B. 2002. Globalization and *Ubuntu*. In Pieter Coetzee and Abraham Roux (eds.), *Philosophy from Africa: A Text with Readings* (2nd ed.), 626–50. Oxford: Oxford University Press.

Raymond, Eric. 2004. The Cathedral and the Bazaar, In Richard A. Spinello and Herman T. Tavani (eds.), *Readings in Cyberethics* (2nd ed.), 367–96. Sudbury, MA: Jones and Bartlett.

Reidenberg, Joel. 2000. Testimony of Joel R. Reidenberg . . . before the Subcommittee on Courts and Intellectual Property Committee on the Judiciary, United States House of Representatives: Oversight Hearing on Privacy and Electronic Commerce, May 18, 2000. <http://reidenberg.home.sprynet.com/Reidenberg_Testimony.html>, accessed December 10, 2007.

Rheingold, Howard. 1990. Teledildonics: Reach Out and Touch Someone. *Mondo 2000*. Issue 2/Summer, 52–4, reprinted in Dennis D. Waskul (ed.), *net.seXXX: Readings on Sex, Pornography, and the Internet*, 319–21. New York: Peter Lang.

Roberds, Stephen C. 2004. Technology, Obscenity, and the Law: A History of Recent Efforts to Regulate Pornography on the Internet. In Dennis D. Waskul (ed.), *net.seXXX: Readings on Sex, Pornography, and the Internet*, 295–316. New York: Peter Lang.

Rose, Flemming. 2006. Why I Published Those Cartoons. *Washington Post*, February 19: B01. <http://www.washingtonpost.com/wp-dyn/content/article/2006/02/17/AR2006021702499_pf.html>, accessed September 30, 2008.

Rosemont, Henry, Jr. 2006. Individual Rights vs. Social Justice: A Confucian Meditation. Lecture, Drury University, April 6.

Ruddick, Sarah. 1989. *Maternal Thinking: Towards a Politics of Peace*. Boston: Beacon.

Saavedra, Marie. 2008. Evidence against Collins Mayor Includes Chat Room Messages, Photos. <http://www.ky3.com/news/local/13755007.html>, accessed September 30, 2008.

Scheule, Rupert M., Rafael Capurro, and Thomas Hausmanninger (eds.)

2004. *Vernetzt gespalten: Der Digital Divide in ethischer Perspektive* [Networked/Split: Ethical Perspectives on the Digital Divide]. (Schriftenreihe des International Center for Information Ethics, Bd. 3). Munich: Wilhelm Fink.

Scollon, Ron and Suzie Wong-Scollon. 2001. *Intercultural Communication* (2nd ed.). Oxford: Blackwell.

Servaes, Jan (ed.). 2007. *Communication for Development and Social Change.* Los Angeles: Sage.

Shelley, Mary Wollstonecraft. [1825] 1933. *Frankenstein: or, a Modern Prometheus.* New York: Dutton.

Shutte, Augustine. 1993. *Philosophy for Africa.* Cape Town: University of Cape Town Press.

Sicart, Miguel. 2005. Game, Player, Ethics: A Virtue Ethics Approach to Computer Games. *International Review of Information Ethics 4* (December). <http://www.i-r-i-e.net/inhalt/ 004/Sicart.pdf>, accessed September 24, 2008.

———. 2009. *The Ethics of Computer Games.* Cambridge, MA: MIT Press.

Silver, James. 2006. "Wave, You're on Catch a Perv!" *The Guardian,* November 16. <http://www.guardian.co.uk/technology/2006/ nov/16/news.g2>, accessed January 16, 2008.

Snapper, John. 2008. The Matter of Plagiarism: What, Why, and If. In Kenneth Einar Himma and Herman T. Tavani (eds.), *The Handbook of Information and Computer Ethics,* 533–52. Hoboken, N.J.: John Wiley and Sons.

Spinello, Richard A. 2008 Intellectual Property: Legal and Moral Challenges of Online File Sharing. In Kenneth Einar Himma and Herman T. Tavani (eds.), *The Handbook of Information and Computer Ethics,* 553–69. Hoboken, N.J.: John Wiley and Sons.

Spinello, Richard A. and Herman T. Tavani. 2004. *Readings in Cyberethics* (2nd ed.). Sudbury, MA: Jones and Bartlett.

Springer, Polly. 1999. Sun on Privacy: "Get Over It." *Wired* (January). <http://www.wired.com/politics/law/news/1999/01/17538>. Accessed December 10, 2007.

Stahl, Bernd Carsten. 2004. *Responsible Management of Information Systems.* Hershey, PA: Idea Group Publishing.

———. 2006. Emancipation in Cross-Cultural IS [Information Systems] Research: The Fine Line between Relativism and Dictatorship of the Intellectual. *Ethics and Information Technology 8* (3), 97–108.

Storsul, Tanja and Dagny Stuedahl (eds.). 2007. *Ambivalence towards Convergence: Digitalization and Media Change.* Gothenburg: NORDICOM.

Story, Louise. 2007. Apologetic, Facebook Changes Ad Program. *New York*

Times, December 6. <http://www.nytimes.com/2007/12/06/ technology/06facebook.html?ei=5070&en=03eaad11b417cc87&ex=11 97694800&adxnnl=1&emc=eta1&adxnnlx=1197118980- V66jhvVdG1DTZaBDnTbeZA>, accessed September 24, 2008.

Stuart, Susan. 2007. Conscious Machines: Memory, Melody and Muscular Imagination. *AI and Consciousness: Theoretical Foundations and Current Approaches*: Papers from the AAAI Fall Symposium, 141–6. Washington, D.C.: Association for the Advancement of Artificial Intelligence.

——. 2008. From Agency to Apperception: Through Kinaesthesia to Cognition and Creation. *Ethics and Information Technology 10* (4), 255–64.

Tavani, Herman T. 2007. *Ethics and Technology: Ethical Issues in an Age of Information and Communication Technology* (2nd ed.). Hoboken, N.J.: John Wiley and Sons.

——. 2008. Floridi's Ontological Theory of Informational Privacy: Some Implications and Challenges. *Ethics and Information Technology 10* (2–3), 155–66.

Thomas, Jim. 2004. Cyberpoaching behind the Keyboard: Uncoupling the Ethics of "Virtual Infidelity." In Dennis D. Waskul (ed.), *net.seXXX: Readings on Sex, Pornography, and the Internet*, 149–77. New York: Peter Lang.

Thomson, Anne. 1999. *Critical Reasoning in Ethics: A Practical Introduction*. London: Routledge.

Time. 1969. Pornography: What is Permitted is Boring. June 6. <http:// www.time.com/time/magazine/article/0,9171,941672–1,00.html>, accessed September 24, 2008.

University of Minnesota Libraries. n.d. Copyright Information and Education: Working with Fair Use. <http://www.lib.umn.edu/ copyright/fairuse.phtml>, accessed September 15, 2008.

van der Velden, Maya. Forthcoming. Design for a Common World: On Ethical Agency and Cognitive Justice. *Ethics and Information Technology*. DOI: 10.1007/s10676-008-9178-2.

Warner, Dorothy E. and Mike Raiter. 2005. Social Context in Massively- Multiplayer Online Games (MMOGs): Ethical Questions in Shared Space. *International Review of Information Ethics 4* (December), 46–52. <http://www.i-r-i-e.net/inhalt/004/Warner-Raiter.pdf>, accessed September 24, 2008.

Warren, Karen J. 1990. The Power and the Promise of Ecological Feminism. *Environmental Ethics 12* (2), 123–46.

Weber, René, Ute Ritterfeld, and Klaus Mathiak. 2006. Does Playing Violent Video Games Induce Aggression? Empirical Evidence of a

Functional Magnetic Resonance Imaging Study. *Media Psychology 8*, 39–60.

Weckert, John. 2007. Giving and Taking Offence in a Global Context. *International Journal of Technology and Human Interaction 3* (3), 25–35.

Weston, Anthony. 2000. *A Rulebook for Arguments* (3rd ed.). Indianapolis, Indiana: Hackett Publishing.

Wheeler, Deborah. 2006. Gender Sensitivity and the Drive for IT: Lessons from the NetCorps Jordan Project. *Ethics and Information Technology 8* (3), 131–42.

White, Amy. 2006. *Virtually Obscene: The Case for an Uncensored Internet*. Jefferson, N.C.: McFarland & Company.

White, Aoife. 2008a. IP Addresses Are Personal Data, E.U. Regulator Says. *Washington Post*, January 22, D1. <http://www.washingtonpost.com/wp-dyn/content/article/2008/01/21/AR2008012101340.html>, accessed September 24, 2008.

——. 2008b. EU: Search Engines under EU Rules. AP (*Wired News Feed*), February 22. <http://news.wired.com/dynamic/stories/E/EU_ONLINE_PRIVACY?SITE=WIRE&SECTION=HOME&TEMPLATE=DEFAULT&CTIME=2008-01-21-16-02-05>, accessed March 1, 2008.

Wiener, Norbert. 1950. *The Human Use of Human Beings: Cybernetics and Society*. Boston: Houghton Mifflin (2nd rev. ed., New York: Doubleday Anchor, 1954).

Wonderly, Monique. 2008. A Humean Approach to Assessing the Moral Significance of Ultra-Violent Video Games. *Ethics and Information Technology 10* (1), 1–10.

Yu, Jiyuan. 2007. *The Ethics of Confucius and Aristotle: Mirrors of Virtue*. New York: Routledge.

Zhang, Yan Bing. 2008. Communication Modes, Asian. In Wolfgang Donsbach (ed.), *International Encyclopedia of Communication*. <http://www.communicationencyclopedia.com/public/>, accessed September 19, 2008.

Index

Abdat, Sjarif 120
abortion 140, 189, 190
Abrahamic religions 60, 112,
177–7, 178, 214–15; *see also*
Christianity, Islam, Judaism
absolutism 5, 20–2, 179, 181,
188–90, 191, 193–4, 195;
copying and distribution 73, 89,
90, 102; of deontology *see*
deontology; privacy 54, 62; sex
and videogames 139–40, 158; *see
also* meta-ethical theory,
pluralism, relativism
accessibility privacy 56–7, 58; *see
also* privacy
Adam, Alison 224
advertising 107–8, 117, 154
affirming the consequent 185
African Information Ethics
conference (Pretoria, 2007) 218,
225
African views 17; copying and
distribution 78–9, 84–6, 87,
101; cross-cultural
communication 125; theoretical
frameworks 196, 217–19, 225;
see also ubuntu
ahimsa (nonviolence – Buddhist
virtue) 178
Ames, Roger 41, 204, 211, 216,
217

analogue media, compared with
digital 9–11, 14
anonymity 5, 31–6, 114, 119, 120
arête (excellence) 125; *see also* virtue
ethics, virtues
Aristotle 25, 51n, 54, 70, 94, 112,
170, 194, 197, 204, 207–10, 212,
216, 217, 219, 221, 223
Asia-Pacific Development
Information Programme
(APDIP) 103
Association for Computing
Machinery (ACM) 168
Association of Internet
Researchers (AoIR) 18–19, 20
atomism 214–15
Attasammapanidhi (self-guidance)
212
autonomy, autonomous self 111,
200, 212; privacy 48–9, 51–2, 63

Barber, Trudy 164
Barnard, Anne 135
Beacon program 1–2, 50
Bentham, Jeremy 172, 173, 175
Berkeley's Software Distribution
(BSD) 97
"Big Brother" 49, 50, 61
binary code 9, 10
BitTorrent 35
blogs 15